Number Sense Interventions

Other products by Nancy C. Jordan, Ed.D., & Nancy Dyson, Ph.D.

Number Sense Screener™ (NSS™) Set K-1,
Research Edition

By Nancy C. Jordan, Ed.D., & Joseph J. Glutting, Ph.D., with Nancy Dyson, Ph.D.

US$89.95 | Stock Number 72261 | 2012 | ISBN 978-1-59857-226-1

Includes a **User's Guide**, a **Stimulus Book** with all the visual stimuli needed to administer the items, an explicit **Quick Script** for accurate administration, and a pack of 25 **Record Sheets**.

**For more information visit
www.brookespublishing.com**

Number Sense Interventions

by

Nancy C. Jordan, Ed.D.
University of Delaware

and

Nancy Dyson, Ph.D.
University of Delaware

Baltimore • London • Sydney

Paul H. Brookes Publishing Co.
Post Office Box 10624
Baltimore, Maryland 21285-0624
USA

www.brookespublishing.com

Copyright © 2014 by Paul H. Brookes Publishing Co., Inc.
All rights reserved.

"Paul H. Brookes Publishing Co." is a registered trademark of
Paul H. Brookes Publishing Co., Inc.

Number Sense Screener™, NSS™, and the following logo are trademarks of Paul H. Brookes Publishing Co., Inc.:
⋮⋮nss

Typeset by Cenveo Publisher Services, Baltimore, Maryland.
Manufactured in the United States of America by
Versa Press, Inc., East Peoria, Illinois.

Cover image © iStockphoto.com/acilo.

Clip art © 2013 Jupiterimages Corporation.

Purchasers of *Number Sense Interventions* are granted permission to download, print, and/or photocopy the Photocopiable Activity Sheets and Materials for educational purposes. None of the forms may be reproduced to generate revenue for any program or individual. *Unauthorized use beyond this privilege is prosecutable under federal law.* You will see the copyright protection notice at the bottom of each photocopiable form.

For further information about the *Number Sense Screener™ (NSS™), K–1, Research Edition*, please contact Brookes Publishing Co. at www.brookespublishing.com or 1-800-638-3775.

Library of Congress Cataloging-in-Publication Data
Jordan, Nancy C., author.
  Number sense interventions / by Nancy C. Jordan, Ed.D., University of Delaware and Nancy Dyson, Ph.D., University of Delaware.
      pages   cm
  Includes bibliographical references and index.
  ISBN 978-1-59857-291-9 — ISBN 1-59857-291-1 (Spiral bound)
  ISBN 978-1-59857-413-5 — ISBN 1-59857-413-2 (EPUB)
  1. Number concept—Study and teaching (Early childhood)—Activity programs—United States.   2. Learning disabled children—Education (Early childhood)—United States.   I. Dyson, Nancy, author.   II. Title.

QA141.15.J67 2014
372.7'049—dc23   2013024744

British Library Cataloguing in Publication data are available from the British Library.

2023 2022 2021 2020 2019

10   9   8   7   6   5   4   3

Contents

About the Photocopiable Activity Sheets and Materials vii
About the Authors viii
Acknowledgments ix

1 Introduction 1
    What Is Number Sense? 1
    Assessing Number Sense 2
    Number Sense Interventions Research that Supports the Lessons 3
    Guidelines for Implementing the Number Sense Interventions
        Lessons 3
    Introduction to Lesson Activities by Skill and Common Core
        State Standards 5
    Incorporating the Lessons into Daily Classroom Life 8
    Number Sense Interventions Activities Organized by Learning
        Goals with Common Core Framing 9
    Materials List 13
    Materials Made from Black-Line Masters 13
    Teacher-Created Materials 14
    Materials that Can Be Found in the Classroom or Purchased 15
    References 15

2 Number Sense Interventions 17
    Lesson 1 17
    Lesson 2 23
    Lesson 3 30
    Lesson 4 36
    Lesson 5 42
    Lesson 6 47
    Lesson 7 52
    Lesson 8 57
    Lesson 9 63
    Lesson 10 68
    Lesson 11 73
    Lesson 12 78
    Lesson 13 83
    Lesson 14 88
    Lesson 15 92
    Lesson 16 96
    Lesson 17 100
    Lesson 18 104
    Lesson 19 109
    Lesson 20 114
    Lesson 21 119
    Lesson 22 124
    Lesson 23 128
    Lesson 24 133
    Extension Partner Activities 137

3 Photocopiable Activity Sheets 145
Lesson 1 Activity Sheet 146
Lesson 2 Activity Sheet 148
Lesson 3 Activity Sheet 150
Lesson 4 Activity Sheet 152
Lesson 5 Activity Sheet 154
Lesson 6 Activity Sheet 156
Lesson 7 Activity Sheet 158
Lesson 8 Activity Sheet 160
Lesson 9 Activity Sheet 162
Lesson 10 Activity Sheet 164
Lesson 11 Activity Sheet 166
Lesson 12 Activity Sheet 168
Lesson 13 Activity Sheet 170
Lesson 14 Activity Sheet 172
Lesson 15 Activity Sheet 174
Lesson 16 Activity Sheet 176
Lesson 17 Activity Sheet 177
Lesson 18 Activity Sheet 178
Lessons 19–24 Activity Sheet 179

4 Photocopiable Materials 181
Cardinality Chart 182
Subitizing Circle Cards 183
Dot Chart for 2 190
Dot Chart for 3 191
Dot Chart for 4 192
Number Sentence Cards 193
Partner Dot Cards 210
Hundreds Chart 211
Five Frames Master 212
Ten Frames Master 213
Decade Cards 214
Unit Cards 220
Bigger/Smaller Cards 223
Teacher Number List 224
Ten Frame Flash Cards 225
Student Number List 226
Activity 18 Hundreds Chart 227
Activity 19 Hundreds Chart 228
Activity 20 Hundreds Chart 229
Activity 21 Hundreds Chart 230
Activity 22 Hundreds Chart 231
Activity 23 Hundreds Chart 232
Activity 24 Hundreds Chart 233

Index 235

About the Photocopiable Activity Sheets and Materials

Purchasers of this book may download, print, and/or photocopy the activity sheets and materials for educational use. These materials are included with the print book and are also available at **www.brookespublishing.com/jordan/nsi/eforms** for both print and e-book buyers.

About the Authors

Nancy C. Jordan, Ed.D., Professor of Education, School of Education, University of Delaware, 206E Willard Hall, Newark, DE 19716

Dr. Jordan is Principal Investigator of the Number Sense Intervention Project (funded by the National Institute of Child Health and Human Development) as well as the Center for Improving Learning of Fractions (funded by the Institute of Educational Sciences, U.S. Department of Education). She is author or coauthor of many articles on mathematics learning difficulties and has recently published articles in *Child Development, Journal of Learning Disabilities, Developmental Science, Developmental Psychology, Journal of Experimental Child Psychology,* and *Journal of Educational Psychology.* Dr. Jordan holds a bachelor's degree from the University of Iowa, where she was awarded Phi Beta Kappa, and a master's degree from Northwestern University. She received her doctoral degree in education from Harvard University and completed a postdoctoral fellowship at the University of Chicago. Before beginning her doctoral studies, she taught elementary school children with special needs. Dr. Jordan served on the Committee on Early Childhood Mathematics of the National Research Council of the National Academies.

Nancy Dyson, Ph.D., School of Education, University of Delaware, 130 Willard Hall, Newark, DE 19716

Dr. Dyson has had an interest in helping children who struggle with mathematics for many years. She received a bachelor's degree in mathematics from the University of Pennsylvania and a master's degree in mathematics education from the University of Pennsylvania's Graduate School of Education. After raising five daughters and serving 30 years as both a teacher and a school director, she returned to school and received a doctoral degree in education with a concentration in mathematics education from the University of Delaware. Since 2009, she has been the project coordinator for the Number Sense Intervention Project (funded by the National Institute of Child Health and Human Development) at the University of Delaware. Dr. Dyson has coauthored articles for the *Journal of Learning Disabilities, Journal of Educational Psychology,* and *Journal of Mathematics Teacher Education* and has spoken extensively on the topic of kindergarten number sense and the effect of early intervention.

Acknowledgments

We are grateful to the children and teachers of the Christina School District in Delaware. We could not have developed these number sense interventions without their extremely generous cooperation. We also thank our wonderful graduate student research assistants, who carried out the interventions during our field trials: Annalisa Alleyne, Molly Blew, Amber Busby, Brenna Hassinger-Das, Casey Irwin, Elizabeth James, Danielle Jansen, Gabrielle Khoury, and Sara Posey. The development of these number sense interventions was supported by the *Eunice Kennedy Shriver* National Institute of Child Health and Human Development (Grant No. R01 HD059170).

CHAPTER 1

Introduction

Early mathematics matters for long-term achievement outcomes. Kindergarten math concepts are powerful predictors of learning outcomes across content areas, independent of cognitive ability and social class (Duncan et al., 2007). Number sense performance and growth, in particular, are highly predictive of math achievement through elementary and middle school (Jordan, Glutting, & Ramineni, 2009; Jordan, Kaplan, Ramineni, & Locuniak, 2009; Locuniak & Jordan, 2008). Although foundational math content covers multiple topics (e.g., number, shapes, space), there is general agreement that number is of primary importance for school (National Research Council [NRC], 2009). Most of the Common Core State Standards (CCSS) for kindergarten (74%) and first grade (67%) concern knowledge of numbers, number relations, and number operations (National Governors Association Center for Best Practices, & Council of Chief State School Officers, 2010). Moreover, learning disabilities in mathematics are characterized by core deficits in number (Butterworth, Varma, & Laurillard, 2011). Research shows that early number competencies can be taught to most young children (Berch, 2005; Berch & Mazzocco, 2007; Dyson, Jordan, & Glutting, 2013; Jordan, Glutting, Dyson, Hassinger-Das, & Irwin, 2012; Starkey & Klein, 2000). Thus, catching mathematics difficulties early and providing evidence-based interventions are key for helping all children achieve success in school mathematics (Geary, Hoard, Byrd-Craven, Nugent, & Numtee, 2005; Gersten, Jordan, & Flojo, 2005).

WHAT IS NUMBER SENSE?

Number sense involves children's knowledge of number, number relations, and number operations (Malofeeva, Day, Saco, Young, & Ciancio, 2004; NRC, 2009). Number sense in preschool and kindergarten generally involves symbolic representations of numbers, both verbal (e.g., number names) and written (e.g., Arabic numerals), in contrast to more fundamental preverbal number knowledge that appears to develop without much input or instruction (Feigenson, Dehaene, & Spelke, 2004; Jordan & Levine, 2009). Children must be able to apply their basic understanding of numbers to a wide range of situations (Carpenter, Hiebert, & Moser, 1983; Gersten et al., 2005; Hiebert, 1984); for example, there can be 4 cookies, 4 names, 4 turns taken, and so forth (Gelman & Gallistel, 1978; NRC, 2009).

Early number competencies follow a developmental progression in pre-K, kindergarten, and first grade (Clements & Sarama, 2007). Children move from acquiring core knowledge of number (e.g., recognizing small quantities and counting to 10), to number relations (e.g., knowing the number before or after another number, determining the larger of 2 numbers), to number operations (e.g., mentally adding and subtracting with small numbers) with the progression repeating as counting skills develop and larger numbers are learned. This progressive cycle represents the three core areas of numbers sense—number, number relations, and number operations (NRC, 2009).

Number

Counting is paramount to mathematical development because it expands quantitative understanding beyond small sets of three or less, which are initially recognized through subitizing or immediate apprehension of the cardinal value of a number (Baroody, 1987; Baroody, Lai, & Mix, 2006; Ginsburg, 1989). Before formal schooling, many children master the count sequence to at least 10 through their everyday experiences. They learn that each object in a collection is counted only once, the count words are always used in the same sequence (i.e., 1, 2, 3), and the last number in the count always denotes the number of objects in the set (i.e., cardinality principle; Gelman & Gallistel, 1978). Children come to understand that they can enumerate any set presented in any configuration as long as they count each object only once in a stable order (Gelman & Gallistel, 1978). Children also must learn to read, write, and understand numerals (NRC, 2009). Linking numerals to quantities through counting helps children see relations between and among numbers (e.g., more than, less than, equal to; Griffin & Case, 1997).

Number Relations

Being able to think about number relations is a key developmental accomplishment (Case & Griffin, 1990; Griffin, 2002, 2004). Typical 4-year-olds can enumerate a set of 4 or 5 objects and answer the question, "How many?" But answering the question, "Which is bigger, 4 or 5?" requires children to integrate counting concepts to make judgments about quantities without physical objects present. This integration typically takes place later in preschool or kindergarten and allows children to begin to make a mental counting line. Understanding number relations is important for learning addition and subtraction. Deficits in number relations are associated with mathematics learning disabilities (Rousselle & Noël, 2007), emphasizing the importance of having these relational structures in place before first grade.

Number Operations

Competence with simple arithmetic is important for success in elementary school mathematics (Jordan, Kaplan, Ramineni, & Locuniak, 2008). These abilities start to develop much earlier, however. Counting is a key strategy for solving addition and subtraction problems. Knowing that the next number in the count sequence is always one more than the preceding number or one less than the following number allows children to use counting to solve problems in the form of $n + 1$ and $n - 1$ (Baroody, Eiland, & Thompson, 2009). Counting also helps children determine exactly how much less or how much more one number is from another (NRC, 2009). For combinations with totals of 10 or less, children can count out each addend on their fingers and then count all of the fingers to get the total. Many children can count on from the first number to get the total (e.g., for 4 + 3, the child counts 5, 6, 7 to get 7) by the end of kindergarten, which is a more efficient approach than counting out both addends (Baroody et al., 2006). Children who adaptively count in kindergarten develop calculation fluency earlier (Jordan et al., 2008).

The fact that numbers can be broken into smaller sets of numbers is another important insight. For example, the number 4 can be broken into 1 and 3, or into 2 and 2. Fuson, Grandau, and Sugiyama (2001) referred to these combinations as "partners" for particular totals. Thinking about different partners for the same sums and also relating them to subtraction (1 + 3 = 4 and 4 − 1 = 3) encourages children to think flexibly. Young children at risk for mathematics difficulties do not make good use of counting and decomposition strategies to help them calculate with totals of 5 or more (Jordan, Kaplan, Oláh, & Locuniak, 2006).

ASSESSING NUMBER SENSE

Jordan and Glutting (2012) developed the *Number Sense Screener™ (NSS™)* based on research of early predictors of mathematics outcomes. The NSS is a measure that not only reliably identifies who will need support in math (Jordan, Glutting, Ramineni, & Watkins, 2010) but also has strong treatment validity (Dyson et al., 2013; Jordan et al., 2012). The content of the NSS covers the core areas of number sense and includes counting, numeral recognition, number comparisons, nonverbal calculations, story problems, and number combinations.

NUMBER SENSE INTERVENTIONS RESEARCH THAT SUPPORTS THE LESSONS

Strong evidence suggests that number sense and the ability to mathematize from everyday experiences can be developed in most children. Randomized controlled trials at the pre-K level have shown positive effects of mathematics interventions that emphasize number (Baroody et al., 2009; Clements & Sarama, 2007, 2008; Dobbs, Doctoroff, & Fisher, 2003; Klein, Starkey, Sarama, Clements, & Iyer, 2008). Jordan and colleagues developed small-group Number Sense Interventions (NSI) for kindergartners who are at risk (Dyson et al., 2013; Jordan et al., 2012). The interventions targeted number, number relations, and number operations—competencies that underlie mathematics difficulties. A randomized study showed that children in the intervention group improved in number sense as well as in general mathematics achievement relative to children in the control groups with moderate to large effect sizes. Controlling for initial mathematical knowledge at pretest, the number sense children performed better than controls on an immediate posttest and a delayed posttest about 2 months later (Jordan et al., 2012). There were moderate to large effect sizes both on a closely aligned NSS (Jordan & Glutting, 2012) as well as a more distal math achievement test. The NSI is directly based on the findings from the scientific work of Jordan, Dyson, and colleagues.

GUIDELINES FOR IMPLEMENTING THE NUMBER SENSE INTERVENTIONS LESSONS

The NSI targets key number sense competencies through explicit instruction for understanding. Activities build on each other, providing scaffolding to help children learn important mathematical relationships that are necessary for transfer to more advanced mathematics tasks. The following guidelines will help the instructor make the most of each lesson.

General Methods

The series of 24 lessons is designed for small groups of children with one instructor. They are designed to keep all children engaged and attentive. Each lesson should take about 30 minutes, but this may vary according to children's needs. Children can make significant gains if the intervention is administered 3 times per week, but daily intervention would be optimal. Each lesson is carefully scripted for easy administration by teachers, special educators, or trained paraprofessionals. Teacher script appears in bold text throughout the lessons. It is helpful to look at the Number Sense Interventions Activities Organized by Learning Goals with Common Core Framing Chart (Chart 1.1) in this manual. Familiarity with the learning goals and the CCSS will help educators make the most of the activities. In addition, the learning goals help teachers tailor the lessons to the individual needs of students.

Group Seating Arrangements

Materials should be presented in a way that is correctly oriented for the children. Do not have children sitting both across and next to you because that will require some to view the materials upside down, which could cause confusion. For example, for number list activities, when the child should be going up the number list, it will appear that he or she is going down the number list. A typical small-group kidney-shaped table works well.

Gestures

It is important to pay close attention to the gestures described in each lesson (e.g., pointing to numbers while counting). Gestures highlight important concepts and provide another means of scaffolding children's skill development. It is helpful for teachers to practice the gestures so that they are natural and do not have to be read from the script during the lesson.

Lesson Materials

All lesson materials can be purchased or made with little expense. Black-line masters for creating the materials are provided with the Photocopiable Materials. Each lesson begins with a list of learning goals along with a list of materials needed for the lesson. It is important to gather and organize the materials before each lesson to save valuable lesson time.

Children who struggle with early mathematics often have had few opportunities to practice writing numbers. Some kindergarten mathematics curricula downplay writing numbers, which may leave these children unprepared for first grade. Therefore, the NSI includes a number writing activity for every lesson. Black-line masters for these activities are provided with the Photocopiable Activity Sheets.

Error-Free Practice and Error Correction

Error-free practice is important for children who have difficulty learning mathematics (Hiebert & Grouws, 2007). Thus, the NSI presents questions to children in an environment that helps them give a correct response. For example, a Cardinality Chart (see the Photocopiable Materials) used in Lessons 1–10 helps children see the +1 relationship between successive numbers. The numbers 1–10 appear across the bottom of the chart, and the appropriate number of circles are above each number. As the children learn the numbers 1–10, there is a visual that connects quantity to numeral and clearly shows the linear relationship between the counting numbers. The instructor asks, "How much is 4 and 1 more?", points to the column of 4 circles on the chart, and gestures 1 up the "steps" of circles to the column above the number 5. Visuals should be used as long as necessary for error-free practice.

Children will still make errors, even though activities are well scaffolded. Each activity has an error correction protocol that should be followed. Children should not go on to another turn until their error has been corrected. For example, if the instructor holds up a card with 4 circles and says, **How many circles?** and the child responds, "3," The instructor should say, **That's a good try, but there are 4 circles.** The instructor then puts down the card, picks it up again, and asks the question again. If a child is regularly giving incorrect responses, then it is probable that he or she has not acquired a prerequisite skill (e.g., recognizing cardinal value of small quantities). The instructor can go back to previous activities in which the child was successful and then gradually move to more challenging activities.

Linear Representations

Linear representations are used throughout the intervention in order to emphasize the linear nature of the counting numbers. Children using the Cardinality Chart in Lessons 1–10 see that each number is 1 more than the number before it and 1 less than the number after it. Lessons 11–24 use a number list that emphasizes the accumulation of quantity as you count. Children using the number list find numbers before and after, perform calculations of +1 and −1, and find missing numbers. Children are also guided to see that numbers are sets that can be broken into smaller sets by using dots that are printed in a line which are then broken into two parts (partners). Finally, these partners are used to create equations, called *number sentences*, and solve story problems. This consistent representation helps children make necessary connections required for problem solving as they begin to use counting strategies for solving addition and subtraction problems. It also allows them to memorize answers to combinations based on relationships among numbers in the count sequence (Baroody et al., 2009).

Vocabulary

The NSI uses carefully chosen vocabulary (e.g., *before, after, plus, minus, equals, bigger, smaller, more, less, altogether*). As new mathematics vocabulary is introduced, the lessons are focused on the meaning of the word with respect to number. For example, *more* and *less* and *bigger* and *smaller* are introduced with activities comparing numbers using the Cardinality Chart. The word *after* is introduced first with Number Recognition Cards and then with $n + 1$ activities on the number list. The word *before* is introduced first with Number Recognition Cards and then with $n - 1$ activities on the number list.

If the classroom curriculum uses a different word for some of the terms used in the NSI, it is best to change the intervention vocabulary to match the classroom curriculum. Using different terminology in different environments may confuse kindergartners.

Formative Assessments and Pacing

Game-style activities can be used as short assessments of the skills or competencies taught that day or on previous days. Teachers can record student responses in whatever way is easiest for their records. A suggested manner is to record only errors. For example, if the correct response is 5 and the child said 4, then 5/4 would be recorded. Nothing is recorded if the response is correct. Thus, it is easy for the instructor to look over assessment results and only attend to those areas where something is recorded.

Although the lessons are written as 24 individual lessons, they are designed to be flexible. Instructors are encouraged to repeat activities or move at a slower (or faster) pace depending on student(s) needs. Activities should not be skipped, however, even if it seems the child has mastered the concept. Review and practice are important for long-term learning.

INTRODUCTION TO LESSON ACTIVITIES BY SKILL AND COMMON CORE STATE STANDARDS

Each lesson covers a variety of skills that support and complement each other. These skills reflect many of those addressed in the CCSS. CCSS abbreviations are listed in parentheses after each activity. A complete list of the CCSS for Mathematics can be found at http://www.corestandards.org.

Oral Counting (Common Core State Standards: K.CC.1, K.CC.2)

Each lesson begins with an oral counting warm-up. Children practice counting to 10, then 20, then 30, then 40 in Lessons 1–10. Each time the children begin counting at 1, which gives repeated practice with the earlier numbers. Children begin counting at 11 instead of 1 at Lesson 11, which saves time and reinforces a counting-on strategy (although instructors might have children continue counting from 1 if more practice in numbers 1–10 is needed). A Hundreds Chart (see the Photocopiable Materials) is displayed during these activities to help children see the numerical patterns in the base 10 numeration system. Subsequent lessons introduce the remaining decade names through counting by tens. Children practice counting higher (50–100), beginning the count at a variety of numbers. The goal is for children to be able to count to 100 with the flexibility to begin counting at any given number.

Number Recognition and Base Ten Understandings (Common Core State Standards: K.CC.1, K.CC.2, K.CC.4, K.CC.5, K.NBT.1)

Magic Number activities build number recognition skills while simultaneously building base ten understandings. Children learn a new number each day—the Magic Number and focus of the day's activities. Although the CCSS only requires base ten understandings for numbers 11–19, the Magic Number activities take these concepts through 100 in order to give kindergarten students who are struggling a head start for first grade. Magic Number activities are as follows.

1. *Cardinality:* Instructors guide children in building numbers using interlocking blocks and the Cardinality Chart. This activity helps children see that each number is 1 more than the number before it and 1 less than the number after it. Numbers 1–5 are built using the same color, and the number 5 is associated with the number of fingers on one hand. Numbers 6–10 are built with 5 blocks of the initial color and n more of another color, showing that these numbers are $5 + x$.

   Children are introduced to the reason our number system is based on 10 (i.e., we each have 10 fingers) during Lesson 10. Subsequent lessons progress through the teens, helping children see these numbers as 10 and so many ones. Rather than building on the Cardinality Chart, the interlocking blocks are put into sticks of 10 (which represent 10 fingers) and a number of single blocks. Figure 1.1 represents the number 13.

   The numeral is built with Decade and Unit Cards as the number is built with blocks. The numeral 13 is built by first showing the numeral 10 (which corresponds to the stick of 10) and then placing the Unit Card 3 over the 0 in 10 while saying, **10 and 3 more is 13.** The Unit Card 3 is then lifted up and the instructor says, **See the 10 hiding under the 3.** This layering of the numerals helps to reinforce the concept that the numbers 11–19 are 10 and so many more (Fuson et al., 2001). See Lessons 11–19 for details of this activity.

   The second set of 10 blocks is made into another stick of 10 when the Magic Number 20 is reached (i.e., 20 is 2 tens). The numbers 21–29 are built as 2 tens and so many ones. Likewise, the numerals 21–29 are built as 20 and so many ones. Children learn to build the numbers 30–99 as so many tens and so many ones in the final lessons, culminating with the number 100. Children begin to anticipate the Magic Number of the day and are excited to share with the instructor what the number will be as they arrive for their small-group time. This activity could easily be adapted for whole class instruction to provide reinforcement for all children.

2. *Sequencing and Number Recognition:* Each day the Magic Number is added to the pile of Number Recognition Cards. First, the cards are laid down in order as the children count. Once the number 11 is added, a new row is begun, modeling the Hundreds Chart. Likewise, the numbers 21–30 form a third row. The cards are shuffled when the sequencing activity is complete and are used for a number recognition game. Instructors can use this game as an opportunity for formative assessment and progress monitoring of each child's number recognition skills. *Note:* The 0 card is not used in the sequencing activities but it is added to the pile for the number recognition game.

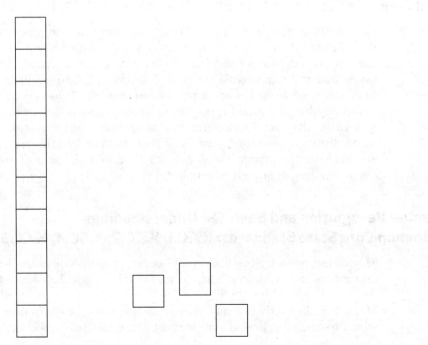

Figure 1.1. Illustration of interlocking blocks to represent the number 13.

3. *Finger Counting:* Children count to the Magic Number on their fingers during Lessons 1–10. Children have several ways of using their fingers to count (e.g., index finger first, pinky first, thumb first). In the NSI, children are encouraged to start with their index finger as this represents the most common model of finger counting. However, if a child is proficient at counting from another finger, do not pressure him or her to change his or her finger-counting habits. Point out that there are 5 fingers on one hand and that numbers 6–10 are 5 and so much more. Children are encouraged to count on from 5 when counting these numbers on their fingers. Ten is shown to be 5 and 5.

Subitizing: Immediate Recognition of Cardinal Values for Small Quantities without Counting (Common Core State Standards: K.CC.6)

Subitizing activities take three forms.

1. *Subitizing Circle Cards:* Children are presented with cards with 0, 1, 2, 3, 4, or 5 circles in progressive lessons. The circles or dots are in varying arrangements. Children are encouraged to say how many dots without counting. When 4 or 5 dots are shown, children are encouraged to look for smaller sets within the set (e.g., 5 dots may show a pattern of 2 dots and 3 dots). The instructor should show the card for only 1 second after the question, "How many?" is asked. Children should not be allowed to count the dots.
2. *Fingers:* Children practice producing and quickly recognizing 1–10 fingers without counting.
3. *Ten Frame Cards:* Children practice recognizing numbers 6–10 without counting by using the patterns inherent in 10 frames. As with the Subitizing Circle Cards, the instructor should show the card for only 1 second after the question, "How many?" is asked.

Partitioning Numbers into Sets: Number Partners (Common Core State Standard: K.OA.3)

Understanding that quantities can be separated into smaller quantities is important in early mathematics. This can be done in more than one way (except for the numbers 1, 2, and 3). For example, 4 can be broken down into 2 + 2 and 3 + 1. The NSI includes several partner activities to teach this concept.

- *Finding Partners:* Children are guided to find all possible ways of breaking a given number into smaller sets.
- *Partners and Number Sentences:* The term *number sentence* is used instead of *equation* in the NSI. Partner Dot Cards provide a visual anchor as children construct number sentences for various number families. Children play matching games in which Partner Dot Cards are matched to Number Sentence Cards.
- *Partners and Story Problems:* Children are asked to match a story problem to a Partner Dot Card. This provides a visual representation, which makes it easier to create the corresponding number sentence.
- *Partners and the Commutative Property:* It is easy to show the commutative property by rotating the partner cards. For example, simply rotating the card 180 degrees shows that both 2 + 3 and 3 + 2 equal 5.

Writing Numerals (Common Core State Standard: K.CC.3)

Each lesson provides an opportunity for children to practice writing numerals. Children initially write the numeral associated with a number of objects and the solutions to simple number sentences (equations). In later lessons, children produce the number sentences associated with a given story problem that they then solve. Instructors may wish to add additional practice in number formation as needed, possibly as a homework assignment.

Using Fingers to Calculate (Common Core State Standard: K.OA.1)

Fingers are always available and can perform the function of a personal calculator. It is important that children learn to efficiently use their fingers, however. Children are first taught to count on their fingers and to subitize small quantities (i.e., children learn to make the number 4 on their fingers quickly, without counting).

Kindergartners should be fluent in combinations for the 1–5 number families and be able to use strategies to solve problems up to the 10 families of operations. Fingers are used in the following ways.

- Children are taught to use their fingers for sums up to 5 to show the part/part/whole model. The sum can then be seen.
- Children are taught a form of counting-on for sums greater than 5.

Totals to 5

Children begin solving addition combinations on their fingers by making one addend on each hand and then counting all of their fingers to find the sum. This method is known as the count-all strategy. For combinations to 5, the NSI has children make one addend on each hand and then to subitize (immediately recognize the cardinal values for small quantities without counting) the answer. Children are trained to subitize a quantity, and it is that quantity that will be presented as the sum of a combination. In this way, children are discouraged from counting quantities they can subitize and are encouraged to quickly state the answer. This is an important bridge for developing fluency.

Totals over 5

The +1 strategy is a natural bridge between counting-all and counting-on. Children first use the Cardinality Chart and then the number list to see that adding 1 results in the next number on the list and subtracting 1 results in the previous number on the list. Adding 1 is the first step in counting-on. As children become comfortable with adding $n + 1$ without first counting n, then they can move to adding 2, then 3, and so forth.

Beginning with the largest number in the combination and counting up the number of the smaller number in the combination is an effective counting strategy (e.g., for the problem 3 + 4, a student would say, "4," and count 5, 6, 7). This strategy allows children to solve combinations more efficiently by counting up rather than counting-all. It also emphasizes cardinality and commutativity. Playing the Great Race Game (Ramani & Siegler, 2008) has been shown to develop these number understandings and may be played as an additional activity.

Story Problems and Number Sentences (Common Core State Standards: K.OA.1, K.OA.2)

Success in primary school mathematics requires children to connect understanding of number transformations to story problems and number combinations. The NSI presents story problems as an application of the number operations learned in the lesson. Children are required to match story problems to Partner Dot Cards and Number Sentence Cards. Next, story problems are modeled using objects, drawings, and tallies. Although drawing a model is effective and leads to correct responses, it is very time consuming. Therefore, children generate their own number sentences in the final lessons to represent and solve a given story problem.

INCORPORATING THE LESSONS INTO DAILY CLASSROOM LIFE

Classroom teachers, special educators, or trained paraprofessionals can administer the lessons in small groups during kindergarten intervention time. They are designed for children who struggle with number concepts or who are at risk for learning difficulties. The structure of the activities encourages children's engagement and development of attention skills. Many of the activities,

Chart 1.1. Number Sense Interventions Activities Organized by Learning Goals with Common Core Framing

Domain		Standard	Lesson 1	2	3	4	5	6	7	8	9	10
Number		Count to 100 by ones and tens.	Count orally to 10	Count orally to 10	Count orally to 10	Count orally to 20	Count orally to 20	Count orally to 20	Count orally to 30	Count orally to 30	Count orally to 30	Count orally to 40
		Count forward beginning from a given number other than 1.						Count to 6 beginning from 5	Count to 7 beginning from 5	Count to 8 beginning from 5	Count to 9 beginning from 5	Count to 10 beginning from 5
		Write numbers from 0 to 20.	Write numbers: 1, 2	Write numbers: 2, 3	Write numbers: 3, 4	Write numbers: 4, 5	Write numbers: 0–5	Write numbers: 5, 6	Write numbers: 6, 7	Write numbers: 7, 8	Write numbers: 8, 9	Write numbers: 9, 10
		Represent a number of objects with a written numeral 0–20.	Connect quantity to numerals 0–2	Connect quantity to numerals 0–3	Connect quantity to numerals 0–4	Connect quantity to numerals 0–5	Connect quantity to numerals 0–5	Connect quantity to numerals 0–6	Connect quantity to numerals 0–7	Connect quantity to numerals 0–8	Connect quantity to numerals 0–9	Connect quantity to numerals 0–10
		When counting objects, say the number names in the standard order, pairing each object with one number (one-to-one correspondence, set enumeration).	Count on fingers to 2	Count on fingers to 3	Count on fingers to 4	Count on fingers to 5	Count on fingers to 5	Count on fingers to 6	Count on fingers to 7	Count on fingers to 8	Count on fingers to 9	Count on fingers to 10
		Understand that the last number said tells the number of objects counted. Count to answer, "How many?" (cardinality).	Count 1,2 objects	Count 1–3 objects	Count 1–4 objects	Count 1–5 objects	Count 1–5 objects	Count 1–6 objects	Count 1–7 objects	Count 1–8 objects	Count 1–9 objects	Count 1–10 objects
		Recognize small quantities 1–5 without counting (subitize).	How many? 1–2	How many? 1–3	How many? 1–4	How many? 1–5	How many? 1–5	How many? 1–6	How many? 1–7	How many? 1–8	How many? 1–9	How many? 1–10
		Recognize quantities 6–10 as 5 + n.	1, 2	1, 2, 3	1, 2, 3, 4	1, 2, 3, 4	1, 2, 3, 4, 5	1, 2, 3, 4, 5	5, 6, 7 Using fingers	5, 6, 7, 8 Using fingers and Ten Frames	5, 6, 7, 8, 9 Using fingers and Ten Frames	5, 6, 7, 8, 9, 10 Using fingers and Ten Frames
	Number relations	Identify whether the number of objects in one group is more than, less than, or equal to the number of objects in another group.					Compare quantities 1–5 more than	Compare quantities 1–5 less than	Compare quantities 1–5 more than/less than	Compare quantities 1–5 more than/less than	Compare quantities 1–5 more than/less than	Compare quantities 1–10 more than/less than
		Compare two numbers between 1 and 10 presented as written numerals or objects.				Before and after 1–5	Before and after 1–5	Before and after 1–6	Before and after 1–7	Before and after 1–8	Before and after 1–9	Before and after 1–10
		Understand that each successive number name refers to a quantity that is one larger.	Build 1 and 2 by adding one more block	Build 1–3 by adding one more block	Build 1–4 by adding one more block	Build 1–5 by adding one more block		Build 1–6 by adding one more block	Build 1–7 by adding one more block	Build 1–8 by adding one more block	Build 1–9 by adding one more block	Build 1–10 by adding one more block
		Compose and decompose numbers from 11 to 19 into tens and ones.										

9

Domain	Standard										
Number operations	Represent and solve addition and subtraction problems with objects, fingers, and equations.	Story problems 2 family	Story problems 3 family	Story problems 4 family (1 + 3, etc.)	Story problems 4 family (2 + 2, etc.)		Story problems 5 family (plus) on Five Frames	Story problems 5 family (minus) on Five Frames	Story problems 5 + (1, 2, 3)	Story problems 5 + (1, 2, 3, 4)	Story problems 5 − (1, 2, 3, 4, 5)
		Number sentences 2 family	Number sentences 3 family	Number sentences 4 family (1 + 3, etc.)	Number sentences 4 family (2 + 2, etc.)	Number sentences 5 family (plus)	Number sentences 5 family (plus)	Number sentences 5 family (minus)	Number sentences 5 + (1, 2, 3)	Number sentences 5 + (1, 2, 3, 4)	Number sentences 5 − (1, 2, 3, 4, 5)
	Decompose numbers less than or equal to 10 into pairs.	Finding partners: 2	Finding partners: 3	Finding partners: 4 (1/3)	Finding partners: 4 (2/2)	Finding partners: 5 on Five Frames			Finding partners: 6, 7, 8 as 5 + n on Ten Frames	Finding partners: 6, 7, 8, 9 as 5 + n on Ten Frames	Finding partners: 6–10 as 5 + n on Ten Frames
	Fluently add and subtract within 5.	Number sentences 1 and 2 families	Number sentences 2 and 3 families	Number sentences 2, 3, 4 (1/3) families	Number sentences 4 (2 + 2, etc.) families	Number sentences 5 family (plus)			Review	Review	Review

Domain	Standard	Lesson									
		11	12	13	14	15	16	17	18	19	20
Number	Count to 100 by ones and tens.	Count orally to 40	Count orally to 40	Count orally to 50	Count orally to 80, by tens to 100	Count orally to 90, by tens to 100	Count orally to 100, by tens to 100	Count orally to 100, by tens to 100	Count orally to 100, by tens to 100	Count orally to 100, by tens to 100	Count orally to 100
	Count forward beginning from a given number.	From 11	From 11	From 21	From 45, 55, 65	From 65, 75, 85	From 76, 86	From 25, 47, 54, 66, 78	From 57, 64, 76, 88	From 25, 67, 74, 86, 95	From 25, 36, 47, 55, 66, 77, 85, 96
		Count to 6–10 beginning from 5	Count to 6–10 beginning from 5	Count to 6–10 beginning from 5	Count to 6–10 beginning from 5	Count to 6–10 beginning from 5	Count to 6–10 beginning from 5	Count to 6–10 beginning from 5	Count to 6–10 beginning from 5	Count to 6–10 beginning from 5	Count on to 6–10 beginning from 5
					Count on to solve sums of 6	Count on to solve sums of 7	Count on to solve sums of 8	Count on to solve sums of 9	Count on to solve sums of 10	Count on using fingers	Count on using fingers
	Write numbers from 0 to 20.	Write number sentences	Write number sentences	Write number sentences	Write number sentences	Write number sentences	Write number sentences	Write number sentences	Write number sentences	Two-digit numbers Write number sentences	Two-digit numbers Write number sentences
	Represent a number of objects with a written numeral 0–20.									Write number sentence for story problem	Write number sentence for story problem
			Count on fingers to 12	Count on fingers to 13	Count on fingers to 14	Count on fingers to 15	Count on fingers to 16	Count on fingers to 17	Count on fingers to 18	Count on fingers to 19	Count on fingers to 20
	When counting objects, say the number names in the standard order, pairing each object with one number (one-to-one correspondence, set enumeration).	Count 1–11 objects	Count 1–12 objects	Count 1–13 objects	Count 1–14 objects	Count 1–15 objects	Count 1–16 objects	Count 1–17 objects	Count 1–18 objects	Count 1–19 objects	Count 1–20 objects

Domain	Objective	How many? 11	How many? 11-12	How many? 11-13	How many? 11-14	How many? 11-15	How many? 11-16	How many? 11-17	How many? 11-18	How many? 11-19	How many? 11-20
Number	Understand that the last number said tells the number of objects counted (cardinality).										
Number	Recognize small quantities 1-5 without counting (subitize).										
Number	Recognize quantities 6-10 as 5 + n.		5-10 using fingers and Ten Frames		5-10 using fingers and Ten Frames	5-10 using fingers and Ten Frames	5-10 using fingers and Ten Frames	5-10 using fingers and Ten Frames	5-10 using fingers and Ten Frames	5-10 using fingers and Penny Ten Frames	5-10 using fingers and Nickel and Penny Ten Frames
Number relations	Identify whether the number of objects in one group is more than, less than, or equal to the number of objects in another group.		Compare quantities 1-10								
Number relations	Compare two numbers between 1 and 10 presented as written numerals.	Bigger/smaller (1-10)		Bigger/smaller (1-10)	Bigger/smaller (1-10)	Bigger/smaller (1-10)		Bigger/smaller (1-10)		Bigger/smaller (1-10)	
Number relations		Before and after 1-10	Before and after 1-10	Before and after 1-10	Before and after 1-10		Before and after 1-10	Before and after 1-10	Before and after 1-10	Before and after 1-10	Before and after 1-10
Number relations	Understand that each successive number name refers to a quantity that is one larger.	After on Number list		Before on Number list							
Number relations		Build 11 as 10 and 1 more	Build 11 and 12 by adding 1	Build 11-13 by adding 1	Build 11-14 by adding 1	Build 11-15 by adding 1	Build 11-16 by adding 1	Build 11-17 by adding 1	Build 11-18 by adding 1	Build 11-19 by adding 1	Build 11-20 by adding 1
Number relations			Plus 1 on number list		Minus 1 on number list	Plus 1, minus 1 on number list	Review game $n + 1/-1$		Review game $n + 1/-1$		Review game $n + 1/-1$
Number operations	Compose and decompose numbers from 11 to 19 into tens and ones.	Build 11 as a stick of 10 plus 1	Build 12 as a stick of 10 plus 2	Build 13 as a stick of 10 plus 3	Build 14 as a stick of 10 plus 4	Build 15 as a stick of 10 plus 5	Build 16 as a stick of 10 plus 6	Build 17 as a stick of 10 plus 7	Build 18 as a stick of 10 plus 8	Build 19 as a stick of 10 plus 9	Build 20 as 2 sticks of 10
Number operations	Represent and solve addition and subtraction problems.	Story problems 5 +/− n	Story problems 5 +/− n	Story problems 5 +/− n			Story problems Drawing models	Story problems Drawing models	Story problems Drawing models	Story problems	Story problems
Number operations		Number sentences 5 +/− n; 2-5 families	Number sentences 5 +/− n; 2-5 families	Number sentences 5 +/− n; 2-5 families	Vertical problems; mixed	Vertical problems; +1/−1				Write number sentences for story problems	Write number sentences for story problems
Number operations	Decompose numbers less than or equal to 10 into pairs.										
Number operations	Fluently add and subtract within 5.	Review	Review	Review	Review	Review	Review	Review	Review	Review	Review

Domain	Standard	Lesson 21	Lesson 22	Lesson 23	Lesson 24
Number	Count to 100 by ones and tens.	Count orally to 100	Count orally to 100	Count orally to 100	Count orally to 100
Number	Count forward beginning from a given number.	From 25, 36, 47, 55, 66, 77, 85, 96	From 25, 36, 47, 55, 66, 77, 85, 96	From 25, 36, 47, 55, 66, 77, 85, 96	From 25, 36, 47, 55, 66, 77, 85, 96
Number	Write numbers from 0 to 20.	Count on to solve sums 7-9	Count on to solve sums 8-10	Count on to solve sums 6-10	Count on to solve sums 6-10
Number		Two-digit numbers Write number sentences	Two-digit numbers Write number sentences	Two-digit numbers Write number sentences	Two-digit numbers Write number sentences
Number	Represent a number of objects with a written numeral 0-20.	Write number sentence for story problem	Write number sentence for story problem	Write number sentence for story problem	Write number sentence for story problem
Number	When counting objects, say the number names in the standard order, pairing each object with one number (one-to-one correspondence, set enumeration).	Count 21 objects			
Number	Understand that the last number said tells the number of objects counted (cardinality).	Solve story problems by finger counting	Solve story problems by finger counting	Solve story problems by finger counting	Solve story problems by finger counting
Number	Recognize small quantities 1-5 without counting (subitize).	Make 1-10 on fingers without counting	Make 1-10 on fingers without counting	Make 1-10 on fingers without counting	Make 1-10 on fingers without counting
Number	Recognize quantities 6-10 without counting as 5 + n.	5-10 using fingers and Nickel and Penny Ten Frames	5-10 using fingers and Nickel and Penny Ten Frames	5-10 using fingers and Nickel and Penny Ten Frames	5-10 using fingers and Nickel and Penny Ten Frames
Number relations	Identify whether the number of objects in one group is more than, less than, or equal to the number of objects in another group.				
Number relations	Compare two numbers between 1 and 10 presented as written numerals.	Bigger/smaller (1-10)	Before and after 1-10	Bigger/smaller (1-10)	Before and after 1-10
Number relations	Understand that each successive number name refers to a quantity that is one larger.	Review game $n + 1/-1$ $n = 1-19$	Review game $n + 1/-1$ $n = 1-20$	Review game $n + 1/-1$ $n = 1-30$	Review game $n + 1/-1$ $n = 1-30$
Number operations	Compose and decompose numbers from 11 to 19 into tens and ones.	Build 21 as 2 sticks of 10 plus 1	Build numbers to 100 as a decade plus a unit	Build numbers to 100 as a decade plus a unit	Build numbers to 100 as a decade plus a unit
Number operations	Represent and solve addition and subtraction problems.	Solve combinations with sums 7-9	Solve combinations with sums 8-10	Solve combinations with sums 6-10	Solve combinations with sums 6-10
Number operations		Write number sentences for story problems	Write number sentences for story problems	Write number sentences for story problems	Write number sentences for story problems
Number operations	Decompose numbers less than or equal to 10 into pairs.				
Number operations	Fluently add and subtract within 5.	Review	Review	Review	Review

however, can also be incorporated in kindergarten classes more generally; they are consistent with the kindergarten CCSS and may be used to supplement kindergarten mathematics curricula. For example, the Cardinality Chart and number list activities would be helpful to most children. The lessons may also be used for pre-K children who are high functioning.

MATERIALS LIST

Many of the activities require using charts, cards, activity sheets, and black-line masters. These resources are referenced in the following materials list and on the first page of the appropriate lesson. In each lesson, representations of the various materials are shown next to the activity script. Teachers should refer to the materials list at the beginning of each lesson to ensure they have the correct materials ready. Black-line masters for some of the materials are included in the book (see Chapters 3 and 4), but other materials might be found in your classroom or can be easily made from inexpensive materials. Directions for making certain materials can be found next.

MATERIALS MADE FROM BLACK-LINE MASTERS

The following cards and charts should be photocopied on card stock and laminated. This will ensure that the materials will last for many lessons. It is easier to cut cards apart *after* laminating. Charts can be made magnetic by adding a magnetic strip to the back, which allows the instructor to display the chart on a magnetic table top easel. To make it easier to sort the materials, feel free to color-code the cards by copying the cards on different colored card stock using pastel colors so the print is easily seen—for example,

Subitizing Circle Cards: Green

Number Sentence Cards: Pink

Partner Dot Cards: Pink (It helps make the connection between number sentences and partners if both sets of cards are the same color.)

Decade Cards: Yellow

Unit Cards: Yellow (These should be the same color as the Decade Cards.)

Bigger/Smaller Cards: Purple

Ten Frame Flash Cards: These work best on a white background.

Cards and Charts

Cardinality Chart
Subitizing Circle Cards
Dot Chart for 2
Dot Chart for 3
Dot Chart for 4
Number Sentence Cards
Partner Dot Cards
Hundreds Chart
Decade Cards
Unit Cards
Bigger/Smaller Cards
Teacher Number List
Ten Frame Flash Cards
Student Number List

Other Materials Made from Black-Line Masters

The following materials should be printed or photocopied on plain copy paper.

Five Frame Mat: The Five Frame Mat should be cut with a 2-inch border at the top and bottom. The mat can be put into a plastic sleeve that has been cut to size. Laminating is not advisable as it makes the mat too thick and the magnets will not stick through a thick layer. This mat can be used as a teacher tool when placed on the magnetic table top easel. It is also used as a student tool when placed in a small metal tray (see Materials that Can Be Found in the Classroom or Purchased).

Ten Frame Mat: Follow the instructions for the Five Frame Mat but leave only a 1-inch border.

Teacher Hundreds Charts for Lessons 18–19

Student Hundreds Charts for Lessons 20–24

Activity Sheets 1–24: The Activity Sheets with two pages should be photocopied so that the Activity Sheet is one double-sided page.

TEACHER-CREATED MATERIALS

Number Recognition Cards: Blank 2-inch by 3-inch playing cards can be purchased in bulk from a variety of stores. These cards work better than traditional flash cards because they are smaller and easier to use for sequencing. Numbers 1–30 should be written on the cards with permanent marker.

Vertical Flash Cards: Any commercially made flash cards can be used, or feel free to make your own.

Two-Colored Dots for Five Frame Mat and Ten Frame Mat: These can be made by purchasing ¾-inch magnetic dots (available at craft stores) and ¾-inch colored dot stickers (available at office supply stores). Simply peel and stick a red dot to one side of the magnet and a yellow dot to the other side. Using strong magnetic dots will keep children from feeling frustrated when their dots slide around, and the teacher can preset the trays and know the dots will stay in place. Commercially available two-colored magnetic dots are not strong enough to stay on the mat.

Penny Magnets for Ten Frame Mat: These can be made by gluing plastic pennies to strong round magnets purchased at a craft store. Physical pennies work better than images of pennies.

Nickel Magnet Strips for Ten Frame Mat: These can be made by gluing a plastic nickel to a card stock strip the size of the top row of the Ten Frame Mat. A magnetic strip should be added to the back of the card stock strip so that it can stick to the mat.

Optional Dime Card: If a dime is to be introduced, then the entire Ten Frame Mat can be covered with a rectangle to which a plastic dime has been glued. Leave a ¼-inch border of the Ten Frame Mat so that children can see it behind the dime. Again, a magnetic strip should be added to the back of the rectangle.

Penny Flash Cards: Penny Flash Cards can be made by gluing plastic pennies to a copy of the Ten Frame Flash Cards (i.e., the pennies cover the black dots). These flash cards can also be made by creating a computer image of the Ten Frame Mat and inserting images of pennies which are easily found on the Internet. These cards must be printed or photocopied in color. (A black and white image of coins would be confusing to children.)

Nickel and Penny Flash Cards: These flash cards can be made using Ten Frame Flash Cards for 5-10 cents. For each card, the nickel should be glued to a strip that almost covers the 5 black dots on the top row. The pennies are glued to the dots on the bottom row. These flash cards can also be made by creating a computer image of the Ten Frame Mat and inserting images of a nickel and pennies which are easily found on the Internet. These cards must be printed or photocopied in color. (A black and white image of coins would be confusing to children.)

MATERIALS THAT CAN BE FOUND IN THE CLASSROOM OR PURCHASED

White board magnetic easel

White board markers

55 interlocking blocks (40 blocks of one color and 15 blocks of another color)

4 pencils

4 erasers

4 crayons

Farm scene with pond and pig pen

5 pig magnets: These can be made by printing small clip art images of a pig on cardstock and gluing them to a magnet.

8 duck magnets: These can be made like the 5 pig magnets.

Metal tray for Five and Ten Frame Mats: Small metal baking pans or trays can be purchased at many discount stores.

Small sticky notes (1½-inch by 2-inches)

Marker

Black dots: Plain black tokens can be purchased online from game web sites.

REFERENCES

Baroody, A.J. (1987). The development of counting strategies for single-digit addition. *Journal for Research in Mathematics Education, 18*(2), 141–157.

Baroody, A.J., Eiland, M., & Thompson, B. (2009). Fostering at-risk preschoolers' number sense. *Early Education and Development, 20*(1), 49.

Baroody, A.J., Lai, M.-L., & Mix, K.S. (2006). The development of young children's early number and operation sense and its implications for early childhood education. In B. Spodek & O. Saracho (Eds.), *Handbook of research on the education of young children* (pp. 187–221). Mahwah, NJ: Lawrence Erlbaum Associates.

Berch, D.B. (2005). Making sense of number sense: Implications for children with mathematical disabilities. *Journal of Learning Disabilities, 38*(4).

Berch, D.B., & Mazzocco, M.M. (Eds.). (2007). *Why is math so hard for some children? The nature and origins of mathematical learning difficulties and disabilities.* Baltimore, MD: Paul H. Brookes Publishing Co.

Butterworth, B., Varma, S., & Laurillard, D. (2011). Dyscalculia: From brain to education. *Science, 332,* 1049–1053.

Carpenter, T.P., Hiebert, J., & Moser, J.M. (1983). The effect of instruction on children's solutions of addition and subtraction word problems. *Educational Studies in Mathematics, 14*(1), 55–72.

Case, R., & Griffin, S. (1990). Child cognitive development: The role of central conceptual structures in the development of scientific and social thought. In E.A. Hauert (Ed.), *Developmental psychology: Cognitive, perceptuo-motor, and neurological perspectives* (pp. 193–230). New York, NY: Elsevier.

Clements, D.H., & Sarama, J. (2007). Effects of a preschool mathematics curriculum: Summative research on the Building Blocks project. *Journal for Research in Mathematics Education, 38,* 136–163.

Clements, D.H., & Sarama, J. (2008). Experimental evaluation of the effects of a research-based preschool mathematics curriculum. *American Educational Research Journal, 45*(2), 443–494.

Dobbs, J., Doctoroff, G.L., & Fisher, P.H. (2003). The "Math is Everywhere" preschool mathematics curriculum and its success in Head Start classrooms. *Teaching Children Mathematics, 10*(1), 20–22.

Duncan, G.J., Dowsett, C.J., Claessens, A., Magnuson, K., Huston, A.C., Klebanov, P., Pagani, ... Japel, C. (2007). School readiness and later achievement. *Developmental Psychology, 43*(6), 1428–1446.

Dyson, N., Jordan, N.C., & Glutting, J. (2013). A number sense intervention for low-income kindergartners at risk for math difficulties. *Journal of Learning Disabilities, 46,* 166–181. doi: 10.1177/0022219411410233

Feigenson, L., Dehaene, S., & Spelke, E. (2004). Core systems of number. *Trends in Cognitive Sciences, 8*(7), 307–314.

Fuson, K.C., Grandau, L., & Sugiyama, P.A. (2001). Achievable numerical understandings for all young children. *Teaching Children Mathematics, 7*(9), 522–526.

Geary, D.C., Hoard, M.K., Byrd-Craven, J., Nugent, L., & Numtee, C. (2005). Early identification and intervention for students with mathematics difficulties. *Journal of Learning Disabilities, 38*(4), 324–332.

Gelman, R., & Gallistel, C.R. (1978). *The child's understanding of number.* Cambridge, MA: Harvard University Press.

Gersten, R., Jordan, N.C., & Flojo, J.R. (2005). Early identification and interventions for students with mathematics difficulties. *Journal of Learning Disabilities, 38*(4), 293–304.

Ginsburg, H.P. (1989). *Children's arithmetic.* Austin, TX: PRO-ED.

Griffin, S. (2002). The development of math competence in the preschool and early school years: Cognitive foundations and instructional strategies. In J.M. Roher (Ed.), *Mathematical cognition* (pp. 1–32). Greenwich, CT: Information Age Publishing.

Griffin, S. (2004). Teaching number sense. *Educational Leadership, 61*(5), 39–42.

Griffin, S., & Case, R. (1997). Re-thinking the primary school math curriculum: An approach based on cognitive science. *Issues in Education, 3*(1), 1–49.

Hiebert, J. (1984). Children's mathematics learning: The struggle to link form and understanding. *Elementary School Journal, 84*(5), 496–513.

Hiebert, J., & Grouws, D.A. (2007). The effects of classroom mathematics teaching on students' learning. In F.K. Lester (Ed.), *Second handbook of research on mathematics teaching and learning* (pp. 371–404). Charlotte, NC: Information Age Publishers.

Jordan, N.C., & Glutting, J.J. (with Dyson, N.). (2012). *The Number Sense Screener™ (NSS™) user's guide, K–1* (Research ed.). Baltimore, MD: Paul H. Brookes Publishing Co.

Jordan, N.C., Glutting, J., Dyson, N., Hassinger-Das, B., & Irwin, C. (2012). Building kindergarteners' number sense: A randomized controlled study. *Journal of Educational Psychology, 104,* 647–660. doi: 10.1037/a0029018

Jordan, N.C., Glutting, J., & Ramineni, C. (2009). The importance of number sense to mathematics achievement in first and third grades. *Learning and Individual Differences, 20*(2), 82–88.

Jordan, N.C., Glutting, J., Ramineni, C., & Watkins, M.W. (2010). Validating a number sense screening tool for use in kindergarten and first grade: Prediction of mathematics proficiency in third grade. *School Psychology Review, 39,* 181–195.

Jordan, N.C., Kaplan, D., Oláh, L., & Locuniak, M.N. (2006). Number sense growth in kindergarten: A longitudinal investigation of children at risk for mathematics difficulties. *Child Development, 77,* 153–175. doi:10.1111/j.1467-8624.2006.00862.x

Jordan, N.C., Kaplan, D., Ramineni, C., & Locuniak, M.N. (2008). Development of number combination skill in the early school years: When do fingers help? *Developmental Science, 11*(5), 662–668.

Jordan, N.C., Kaplan, D., Ramineni, C., & Locuniak, M.N. (2009). Early math matters: Kindergarten number competence and later mathematics outcomes. *Developmental Psychology, 45*(3), 850–867.

Jordan, N.C., & Levine, S.C. (2009). Socioeconomic variation, number competence, and mathematics learning difficulties in young children. *Developmental Disabilities Research Reviews, 15*(1), 60–68.

Klein, A., Starkey, P., Sarama, J., Clements, D.H., & Iyer, R. (2008). Effects of a pre-kindergarten mathematics intervention: A randomized experiment. *Journal of Research on Educational Effectiveness, 1,* 155–178.

Locuniak, M.N., & Jordan, N.C. (2008). Using kindergarten number sense to predict calculation fluency in second grade. *Journal of Learning Disabilities, 41*(5), 451–459.

Malofeeva, E., Day, J., Saco, X., Young L., & Ciancio, D. (2004). Construction and evaluation of a number sense test with Head Start children. *Journal of Educational Psychology, 96*(4), 648–659.

National Governors Association Center for Best Practices, & Council of Chief State School Officers. (2010). *Common Core State Standards for mathematics.* Retrieved from http://www.corestandards.org/assets/CCSSI_Math%20Standards.pdf

National Research Council. (2009). *Mathematics learning in early childhood: Paths toward excellence and equity.* Washington, DC: National Academies Press.

Ramani, G.B., & Siegler, R.S. (2008). Promoting broad and stable improvements in low-income children's numerical knowledge through playing number board games. *Child Development, 7*(2), 375–394.

Rousselle, L., & Noël, M.-P. (2007). Basic numerical skills in children with mathematics learning disabilities: A comparison of symbolic vs. non-symbolic number magnitude processing. *Cognition, 102,* 361–395.

Starkey, P., & Klein, A. (2000). Fostering parental support for children's mathematical development: An intervention with Head Start families. *Early Education and Development, 11*(5), 659–680.

CHAPTER 2

Number Sense Interventions

Lesson 1

Learning Goals

Establish behavior boundaries
Count to 10 orally
Build numbers 1, 2 using blocks
Number recognition 0–2
Count and sequence to 2
Count to 2 on fingers
Make numbers 1, 2 on fingers
Recognize quantities 1, 2
Partners of 2 (1/1)
Story problems and number sentences: 2 family
Perform number operations on fingers
Write the numerals 0–2
Connect the quantity to numerals 0–2
Solve written number sentences: 1 and 2 families

Materials

COPY

Cardinality Chart
Subitizing Circle Cards (#1–12)
Dot Chart for 2
Partner Dot Card for 2
Number Sentence Cards: 1 + 1 = 2, 2 − 1 = 1, 1 − 1 = 0, 2 − 2 = 0
Lesson 1 Activity Sheet

GATHER

White board magnetic easel
3 interlocking blocks in a plastic bag
2 of each: pencils, erasers, crayons, and blocks
Pencils without erasers and crayons

PREPARE

Put student names on Lesson 1 Activity Sheets.
Number Recognition Cards (0–2); *See Chapter 1 for instructions on making Number Recognition Cards.*

★ *Teacher Tip:* It is important to gather and organize materials before each lesson to save valuable time.

ESTABLISHING BEHAVIOR BOUNDARIES

- Say, **When we work together, I expect you all to carefully listen and do your best. Okay? When I am talking, you should be listening. And when I call on one of you to speak, we will all listen to you. That is how we show respect for each other. It is important that we respect each other so we can all do our best.**

- **To show me you are ready to learn, your body should be still, sitting just like this, feet under the table, bottom on the chair, and hands to myself.** Wait for imitation. **Exactly! That shows me you are ready.**

COUNTING WARM-UP

- Say, **Let's count to 10. Use your inside voices and let's stay together. Ready? 1, 2, 3, 4, … 10.** Count slowly and clearly.

- Say, **Let's count around the group this time. Each of you will say the next number when it is your turn. I will start ... 1.** Point to the child to your right. Take a turn yourself.
- If a child says the incorrect number, then say, **That was a good try but the next number is ___. Let's try again.** For each error, back up 2 children and repeat so the child has an opportunity to be successful.

MAGIC NUMBER ACTIVITIES (Magic Number is 2)

★ Cardinality

MATERIALS: Cardinality Chart, 3 interlocking blocks in a plastic bag

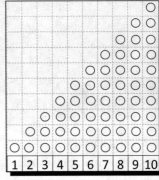

Cardinality Chart [number chart]

- Lay the Cardinality Chart on the table and say, **This is a number chart. It has the numbers 1–10 across the bottom in a list.** Run your finger across the numbers.
- Point to the number 1 and say, **The first number is 1.** Hold up 1 block and say, **This is 1 block. How many blocks do I have?** Wait for a response.
- Put 1 block on the first block above the number 1 and say, **1 block.**
- Point to the number 2 and say, **The next number is 2.** Hold up 2 blocks—one in each hand—and say, **These are 2 blocks. 1, 2. How many blocks do I have?** Wait for a response.
- Put the blocks together and say, **2 blocks.** Then put them right above the number 2 on the chart.
- Say, **As we go up the list of numbers on the number chart** (gesture up the number list on the Cardinality Chart), **we add 1 block at a time. It is like climbing up steps, 1 at a time. See, 1** (put your finger on the block above 1) **and 1 more** (move to the top block above 2) **is 2.** Touch the number 2 and run your finger up the 2 blocks.
- Say, **Let's try that again! 1 and 1 more is ___.** If the children do not answer right away, then use the previous gestures to scaffold the answer.

★ Sequencing and Number Recognition 0–2

MATERIALS: Number Recognition Cards 0–2 (*the 0 card is used for number recognition only—not sequencing*), white board magnetic easel

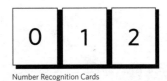

Number Recognition Cards

- Say, **I have cards here with numbers on them. Say the numbers with me as I put them down. Do not go ahead of me. Ready? 1, 2!** Put the cards down horizontally as if building a number list. Orient the cards toward the children.
- Hold up the Number Recognition Card 1 and say, **What number is this?** Wait for a response.
- Hold up the 2 card and say, **What number is this?** Wait for a response.
- Hold up the 0 card and say, **What number is this?** Wait for a response.
- Say, **That's right, this is the number 0. Can you show me 0 blocks?** Children should respond with showing empty hands or saying, "No blocks" or "Nothing."
- Say, **Each day we are going to have a Magic Number. Today our Magic Number is 2.** Hold up the Number Recognition Card 2 and say, **What number is this?** Wait for a response.
- Say, **That is right. This is the number 2. Our Magic Number today is 2.**
- Say, **This is how we write a 2. Watch carefully. Around and across.** Write a 2 on the white board magnetic easel (no loop at the bottom of the numeral 2).
- Say, **Now it is your turn! We are going to write a big number 2 in the air. Follow my finger. Ready? Copy me!**

Number Sense Interventions Lesson 1   19

- Trace a number 2 in the air backward (about 18" high) so that the children will see it in the correct orientation. Say, **Let's try that again!** Repeat twice.
- Say, **Let's play a game. I will hold up a card with a number on it, and I want you to tell me the name of the number. I will point to you when it is your turn, okay? Everyone else, say it in your mind, not out loud.**
- Say, **But, if the Magic Number comes up, I want everyone to say the number, even if it is not your turn. Pay close attention so you will know when our Magic Number comes up!**
- Go around the group, showing each one a different number from 0 to 2 (not in order). Make sure all children can see the number. Go around the group three times.

★ Error Correction

- If the children do not *all* answer for the Magic Number, then say, **This is the Magic Number, so everyone is supposed to answer. Let's try that again!** Hold up the number 2 and say, **What number is this?**
- If a child incorrectly answers, then say, **That is a good try but the number is ___.** Put the card down, hold it up again, and say, **What number is this?**

★ Finger Counting

- See Chapter 1 for a discussion of counting styles.
- Say, **Let's count to 2 on our fingers. Watch me. 1, 2.** Demonstrate starting with index finger.
- Say, **Now you try! 1, 2.** Make sure all the children are correctly counting. **Let's try again. 1, 2.**
- Say, **Now let's try on your other hand. 1, 2. Let's try again. 1, 2.**

SUBITIZING QUANTITIES (0-2) ACTIVITIES

★ Finger Automaticity

- Say, **Let's practice making numbers on our fingers quickly. Let's start with 1. This is how we make 1.** Put up an index finger. **Everyone show me 1 on your fingers.**
- Make sure everyone shows his or her index finger.
- Say, **This is how we make 2.** Show 2 on your fingers. **Everyone show me 2 on your fingers.** Correct any errors.
- Say, **Now we are going to play a game. When I say a number, you hold up that many fingers.** Say the numbers 1 or 2 in random order. Practice until all children are proficient (quick), but no more than 10 trials.

★ Recognizing Sets

MATERIALS: Subitizing Circle Cards #1-12

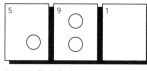
Subitizing Circle Cards

- Say, **Now I am going to hold up a card with some circles on it, and I want you to tell me how many circles are on the card. Try to tell me as fast as you can without counting. Let's try some together.**
- Hold up Subitizing Circle Card #5 (with 1 circle) and say, **How many circles?** After 1 second, put the card down. Use error correction previously listed if needed.
- Say, **Let's try another one!** Hold up card #9 (with 2 circles) and say, **How many circles?** After 1 second, put the card down. Use error correction if needed.
- Hold up card #1 (with 0 circles) and say, **How many circles?** After 1 second, put the card down. Children may say "nothing" or "no circles." If so, then ask, **What number means nothing or no circles?**

- Say, **Now we are going to play a game. Each of you will take a turn telling how many circles are on the card. If it is not your turn, then say the answer in your minds, not out loud.**
- Go around the group three times, showing each child a different card using the following protocol.
  - Hold up the card and say, **How many?** Wait 1 second. Put the card down.
  - If the child does not respond or shows indications of counting, then put the card down and say, **Try to tell me right away—just by looking, not counting.** Hold it up again and say, **How many?** Wait 1 second. Put the card down.
  - If the child responds incorrectly, then say, **That is a good try, but there are __ circles.** Put the card down, hold it up, and say, **How many?** Wait 1 second. Put the card down.

PARTNERS OF 2 ACTIVITIES

MATERIALS: Pencil, Partner Dot Cards for 2, Dot Chart for 2, Number Sentence Cards

Note: Do not use the words *plus* or *equals* until they are introduced in the activity.

★ Partners of 2 (1 + 1) Using Dot Chart for 2

- Put down the Dot Chart with 2 dots. Orient it horizontally.
- Say, **See, here are 2 dots. 1, 2.** Touch the dots as you count. **I can break them into 2 parts with my pencil.** Separate the 2 dots using a pencil.
- Say, **1 and 1 is 2.** Point to the dots when you say, "1 and 1" and circle around the 2 dots when you say, "2."
- Say, **We call 1 and 1 partners for 2 because together they make 2. I made a partner card so I do not have to use my pencil to make parts. See how it looks the same.** Put the Partner Dot Card for 2 above the Dot Chart for 2. Orient it horizontally. **1 and 1 is 2.** Remove the chart and pencil.

Dot Chart for 2

Partner Dot Card for 2

★ Partners and Number Sentences

1 + 1 = 2

- Put down the 1 + 1 = 2 card and say, **Here is another way to show 1 and 1. This is a number sentence. We call this** *plus.* **It means to put these two amounts together. 1 and 1 or 1 plus 1.** Gesture to each addend as you say it.
- Point to the equals sign and say, **We call this** *equals.* **Equals means the** *same amount as.* **1 plus 1 is the same amount as 2.** Gesture to each part of the number sentence as you say it. **1 plus 1 equals 2.**
- Point to the plus sign and say, **What do we call this?** Repeat.
- Point to the equals sign and say, **What do we call this?** Repeat.
- Say, **Let's say this number sentence together: 1 plus 1 equals 2.**
- Put down the Partner Dot Card for 2 and say, **See how the dots on each part match the numbers on the plus card. 1 plus 1 equals 2.** Point to each dot, and then circle the 2 dots.
- Say, **Let's say it again. 1 plus 1 equals 2.** Touch the numbers and the signs on the Number Sentence Card.

Number Sentence Card

Partner Dot Card for 2

2 − 1 = 1 and 2 − 2 = 0

- Using the Partner Dot Card for 2, say, **I can use this partner card to show a take-away problem, too. Watch what I do.**
- Say, **I have 2, but if I take away 1** (cover up the dot on the left), **then I have 1 left.** Point to the remaining dot. Demonstrate again.

Partner Dot Card for 2

Number Sense Interventions Lesson 1   21

- Put down the 2 − 1 = 1 card and say, **Here is another way to show 2 take away 1. We call this minus.** Point to the minus sign. **What do we call this?** Wait for a response. **Minus means take away.**

  2 − 1 = 1
  Number Sentence Card

- Say, **So this says 2 minus 1 equals 1. Let's say this number sentence together. 2 minus 1 equals 1.** Repeat. Touch the numbers and the signs in the number sentence.
- Say, **That means that 2 take away 1 is the same amount as 1.**
- Put down the 2 − 2 = 0 card and using the Partner Dot Card for 2 say, **If I have 2 dots** (circle the 2 dots) **and take away 2 dots** (cover both dots), **how many dots are left?** Wait for a response.

  2 − 2 = 0
  Number Sentence Card

- Say, **So 2 minus 2 equals 0.** Repeat. Touch the numbers and the signs on the Number Sentence Card.

★ Review

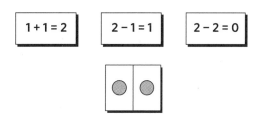

- Put the 1 + 1 = 2, 2 − 1 = 1, and the 2 − 2 = 0 cards next to each other above the Partner Dot Card for 2.
- Say, **Let's say these 1 more time. 1 plus 1 equals 2, 2 minus 1 equals 1, and 2 minus 2 equals 0.** Gesture using the Partner Dot Card for 2 as you say each number sentence.

STORY PROBLEMS

MATERIALS: 2 of each: pencils, erasers, crayons, blocks; Number Sentence Cards

- Say, **Now I am going to tell you a story.** Give a pencil to a child sitting nicely and say, **_(Child)_ has 1 pencil. His/her teacher gave him/her 1 more pencil** (give the child another pencil). **How many pencils does _(child)_ have now?**
- Hold up the 1 + 1 = 2 card and say, **That is right because 1 pencil plus 1 pencil equals 2 pencils.** Point to the numbers as you say them.

  1 + 1 = 2
  Number Sentence Card

- Say, **This story is going to be a little different.** Give another child an eraser and say, **_(Child)_ has 1 eraser. His/her teacher takes the 1 eraser** (take the eraser). **How many erasers does he/she have now?**
- Children may say "nothing" or "no erasers". If so, then ask, **What number means nothing or no erasers?**
- Hold up the 1 − 1 = 0 card and say, **See, this says that 1 eraser minus 1 eraser equals 0 erasers.** Point to the 0 and say, **What number is this?**

  1 − 1 = 0
  Number Sentence Card

- Say, **Here is another story.** Give another child 2 crayons and say, **_(Child)_ has 2 crayons. The teacher takes 1 of his/her crayons** (take 1 crayon). **How many crayons does _(child)_ have now?**
- Hold up the 2 − 1 = 1 card and say, **That is right because 2 crayons minus 1 crayon equals 1 crayon.**

  2 − 1 = 1
  Number Sentence Card

- Say, **Here is 1 more story.** Give another child 2 blocks and say, **_(Child)_ has 2 blocks. The teacher takes the 2 blocks** (take 2 blocks). **How many blocks does _(child)_ have now?** Children may say "nothing" or "no blocks". If so, then ask, **What number means nothing or no blocks?**

- Hold up the 2 − 2 = 0 card and say, **See, this says 2 blocks minus 2 blocks equals 0 blocks.**
- Point to the 0 and say, **What number is this?**

`2 − 2 = 0`
Number Sentence Card

NUMBER SENTENCES ON FINGERS

MATERIALS: Number Sentence Cards

Note: See Chapter 1, Finger Counting, for a trajectory of solving number sentences on fingers.

1 + 1 = 2

- Hold up the 1 + 1 = 2 card and say, **Watch me do this using my fingers.**
- Say, **The first number is 1. This is 1 finger.** Hold up your index finger on your right hand.
- Point to the + 1 on the card and say, **This says + 1 so I will put up 1 more finger.** Put up 1 finger on the left hand.
- Say, **How many fingers do I have up? That is right. 1 plus 1 equals 2.** Emphasize each finger as you say, "1 plus 1" and both together as you say, "equals 2."
- Say, **Now you try it. 1** (check all hands) **plus 1** (check all hands) **equals 2.** Repeat.

`1 + 1 = 2`
Number Sentence Card

2 − 1 = 1

- Put the 2 − 1 = 1 card in front of you. Say, **Watch me do this on my fingers.**
- Say, **The first number is 2. I will put up 2 fingers.** Put 2 fingers up at once, not one at a time.
- Say, **Now I will put them on the table like this.** Put 2 fingers on the edge of the table.
- Say, **This says take away 1 or minus 1. Watch how I take away 1 finger!** Demonstrate holding 1 finger with your other hand, covering it.
- Say, **How many fingers are not covered? That is right. 2 minus 1 equals 1.** Demonstrate again as you speak.
- Say, **Now you try it. 2** (check all hands) **minus 1** (check all hands) **equals 1.** Repeat.

`2 − 1 = 1`
Number Sentence Card

WRITTEN NUMBERS

MATERIALS: Lesson 1 Activity Sheets, pencils, crayons

- Say, **Now we are going to practice writing numbers. Here is a paper with the numbers 0, 1, and 2 on it. I would like you to trace over the numbers with your pencil. Do your best to stay on the dotted lines, but do not erase if you make a mistake. Just try it again. Under the dotted numbers is a place to copy the numbers on your own. Then write the number that tells us how many things are in the box.** Point to the pictures at the bottom of the page.
- As each child finishes, have him or her turn the paper over and say, **Trace the answer for these number sentences. Figure them out on your fingers and see if you get the same answer!** If some children finish early, then they can color the objects on the paper. Have children self-correct their papers by orally counting the objects. Do not let a child color an incorrect paper.

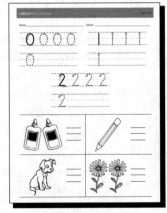
Lesson 1 Activity Sheet

Number Sense Interventions Lesson 2   23

Lesson 2

Learning Goals	Materials
Establish behavior boundaries	**COPY**
Count to 10 orally	Cardinality Chart
Build numbers 1–3 using blocks	Subitizing Circle Cards (#1–18)
Number recognition 0–3	Dot Chart for 3
Count and sequence to 3	Partner Dot Card for 3
Count to 3 on fingers	Number Sentence Cards: 2 + 1 = 3, 1 + 2 = 3, 3 − 1 = 2, 3 − 2 = 1,
Make numbers 1–3 on fingers	3 − 3 = 0
Recognize quantities 0–3	Lesson 2 Activity Sheet
Partners of 3 (1/2)	**GATHER**
Story problems and number sentences: 3 family	White board magnetic easel
Perform number operations on fingers	6 interlocking blocks in a plastic bag
Write the numerals 2, 3	3 of each: pencils, erasers, crayons, and blocks
Connect the quantity to numerals 0–3	Pencils without erasers and crayons
Solve written number sentences: 2 and 3 families	**PREPARE**
	Put student names on Lesson 2 Activity Sheets.
	Number Recognition Cards (0–3); *See Chapter 1 for instructions on making Number Recognition Cards.*

ESTABLISHING BEHAVIOR BOUNDARIES

- Say, **When we work together, I expect you all to carefully listen and do your best. Okay? When I am talking, you should be listening. And when I call on one of you to speak, we will all listen to you. That is how we show respect for each other. It is important that we respect each other so we can all do our best.**

- **To show me you are ready to learn, your body should be still, sitting just like this, feet under the table, bottom on the chair, and hands to myself.** Wait for imitation. **Exactly! That shows me you are ready.**

COUNTING WARM-UP

- Say, **Let's count to 10 taking turns like last time. Each of you will say the next number when it is your turn. I'll start. 1.** Point to the child to your left. If a child says the incorrect number, then say, **That was a good try but the next number is __. Let's try again.** For each error, back up 2 children and repeat so the child has an opportunity to be successful. Take a turn yourself.

MAGIC NUMBER ACTIVITIES (Magic Number is 3)

★ Cardinality

MATERIALS: Cardinality Chart, 6 interlocking blocks in a plastic bag

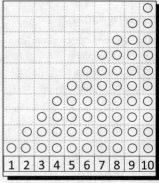

Cardinality Chart [number chart]

- Lay the Cardinality Chart on the table and say, **Here is our number chart. It has the numbers 1–10 across the bottom in a list.** Run your finger across the numbers.
- Point to the number 1 and say, **The first number is 1.** Hold up 1 block and say, **This is 1 block. How many blocks do I have?** Wait for a response.
- Put 1 block on the first block above the number 1 and say, **1 block.**
- Point to the number 2 and say, **The next number is 2.** Hold up 2 blocks—one in each hand—and say, **These are 2 blocks. 1, 2. How many blocks do I have?** Wait for a response.
- Put the blocks together and say, **2 blocks.** Then put them right above the number 2 on the chart.
- Point to the number 3 and say, **The next number is 3.**
- Put 2 more blocks together and put them in one hand. Put 1 block in your other hand and say, **2 and 1 more is 3.** Put the blocks together and put them on the chart above the 3 and say, **3 blocks.**
- Say, **As we go up the number list** (gesture up the number list on the Cardinality Chart), **we add 1 block at a time. It is like climbing up steps, 1 at a time. See, 1** (put your finger on the block above 1) **and 1 more** (move to the top block above 2) **is 2, and 2 and 1 more is 3.** Touch the number 3 and run your finger up the 3 blocks.
- Say, **Let's try that again! 1 and 1 more is ___.** If the children do not answer right away, then use the previous gestures to scaffold the answer.

★ Sequencing and Number Recognition 0–3

MATERIALS: Number Recognition Cards 0–3, white board magnetic easel

Number Recognition Cards

- Pick up the Number Recognition Cards, add the 0 card, and shuffle the cards for the number recognition game.
- Say, **I have cards here with some numbers written on them. Say the numbers with me as I put them down. Do not go ahead of me. Ready? 1, 2, 3!** Put the cards down horizontally as if building a number list. Orient the cards toward the children.
- Hold up the Number Recognition Card 3 and say, **What number is this?** Wait for a response.
- Say, **That is right. This is the number 3. Our Magic Number today is 3.**
- Say, **This is how we write a 3. Watch carefully—around and around.** Write a 3 on the white board magnetic easel.
- Say, **Now it is your turn! We are going to write a big number 3 in the air. Follow my finger. Ready? Copy me!**
- Trace a number 3 in the air backward (about 18" high) so that the children will see it in the correct orientation. Say, **Let's try that again!** Repeat twice.
- Say, **We are going to play a game. I will hold up a card with a number on it, and I want you to tell me the name of the number. I will point to you for your turn, okay? Everyone else, say it in your mind, not out loud.**
- Say, **But, if the Magic Number comes up, I want everyone to say the number, even if it is not your turn. What is the Magic Number for today?** Wait for a response. **Pay close attention so you will know when our Magic Number comes up!**

Number Sense Interventions Lesson 2

- Go around the group, showing each child a different number from 0 to 3 (not in order). Make sure all children can see the number. Go around the group three times.

★ Error Correction

- If they do not all answer for the Magic Number, then say, **This is the Magic Number. Everyone is supposed to answer. Let's try that again!** Hold up the number 3 and say, **What number is this?**
- If a child incorrectly answers, then say, **That is a good try but the number is ___.** Put the card down, hold it up again, and say, **What number is this?**

★ Finger Counting

- Say, **Let's count to 3 on our fingers. Watch me. 1, 2, 3.** Demonstrate starting with index finger.
- Say, **Now you try! 1, 2, 3.** Make sure all the children are correctly counting. **Let's try again. 1, 2, 3.**

SUBITIZING QUANTITIES (1–3) ACTIVITIES

★ Finger Automaticity

MATERIALS: Number Recognition Cards 1–3

- Say, **Let's practice making numbers on our fingers quickly. Let's start with 1. This is how we make 1.** Put up an index finger. **Everyone show me 1 on your fingers.** Make sure everyone shows his or her index finger.
- Say, **This is how we make 2.** Show 2 on your fingers. **Everyone show me 2 on your fingers.** Correct any errors.
- Say, **This is how we make 3.** Show 3 on your fingers. **Everyone show me 3 on your fingers.** Correct any errors.
- Say, **Today we are going to play a new game. I want you to look at the number I show you and put up that many fingers. If I show you this card** (show the Number Recognition Card 1), **how many fingers should you put up?** Wait for a response.

Number Recognition Cards

- Say, **That is right. Show me 1 finger.** Make sure everyone shows his or her index finger.
- Repeat with 2 and 3 fingers.
- Shuffle the Number Recognition Cards and show them one at a time. Go through the cards four times, shuffling each time.
- Be sure that each child is holding up the correct number of fingers. If he or she is incorrect, then have him or her look at all the other children and correct him- or herself or let another child help him or her. Discourage counting and encourage automaticity.

★ Recognizing Sets

MATERIALS: Subitizing Circle Cards #1–18

Subitizing Circle Card

- Hold up Subitizing Circle Card #13 (with 3 circles) and say, **Here is a card with 3 circles on it.** Put it down and hold it up. **How many circles?** Wait for a response.
- Repeat with Subitizing Circle Cards #14–18.
- Shuffle the new cards into the pile toward the front.

- Say, **Now I am going to hold up a card with some circles on it, and I want you to tell me how many circles are on the card. Try to tell me as fast as you can without counting. I will point to you for your turn, okay? Everyone else, say it in your mind.**
- Go through all the Subitizing Circle Cards in random order. Use protocol from Lesson 1.

PARTNERS OF 3 ACTIVITIES

MATERIALS: Pencil, Partner Dot Card for 3, Number Sentence Cards, Dot Chart for 3

★ Partners of 3 (2 + 1) Using Dot Chart for 3

Dot Chart for 3

- Put down the Dot Chart for 3. Orient it horizontally.
- Say, **See, here are 3 dots. 1, 2, 3.** Touch the dots as you count. **I can break them into 2 parts with my pencil.** Make sure the 2 dots part is at the child's left (your right).
- Say, **2 dots plus 1 dot equals 3 dots.** Circle dots on either side of the pencil as you say, "2 plus 1" and circle all the dots when you say, "3."
- Put down the Partner Dot Card for 3, oriented like the Dot Chart. Gesture to each set of dots as you say, **See, 3 has 2 and 1 hiding inside it.**

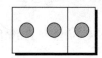
Partner Dot Card for 3

- Say, **2 plus 1 equals 3.** Point to the sets of dots as you say, "2 plus 1" and circle the 3 dots as you say, "equals 3."
- Say, **How much is 2 plus 1?** Continue to gesture to the two sides of the card and then circle all the dots. Repeat.

★ Partners of 3 (1 + 2) Using Dot Chart for 3

- Say, **Let's move the pencil to another place.** Move the pencil over 1 dot. Point to the 1 dot side and say, **How many dots are in this part?** Point to the 2 dot side and say, **How many dots are in this part?**
- Say, **How many dots are there in all?** Wait for a response. Say, **That is right! 1 dot plus 2 dots equals 3 dots.** Circle dots on either side of the pencil as you say, "1 plus 2" and circle all the dots when you say, "3."
- Point to the Partner Dot Card for 3 and say, **How can I make this card show 1 plus 2 instead of 2 plus 1?**
- If no one responds, then rotate the card 180 degrees.
- Say, **Now this partner card shows 1 plus 2 equals 3. See, 3 has 1 and 2 hiding inside it.** Continue gesturing to the sets of dots.
- Say, **How much is 1 plus 2?** Continue to gesture to the 2 sides of the Partner Dot Card and then circle all the dots. Repeat.

★ Partners and Number Sentences

2 + 1 = 3

Number Sentence Card

- Rotate the Partner Dot Card so it shows 2 + 1.
- Put down the 2 + 1 = 3 card and say, **Here is another way to show 2 plus 1.**
- Say, **This says 2 plus 1 equals 3.**
- Say, **That means that 2 plus 1 is the same amount as 3.**

Partner Dot Card for 3

Number Sense Interventions Lesson 2

- Say, **See how the dots on each part match the numbers on the plus card. 2 plus 1 equals 3.** Gesture the quantities.
- Say, **Let's say this number sentence together. 2 plus 1 equals 3.** Repeat. Touch the numbers and the signs on the 2 + 1 = 3 card.

1 + 2 = 3

| 1 + 2 = 3 |
| Number Sentence Card |

- Rotate the Partner Dot Card 180 degrees and say, **Now our partner card shows 1 plus 2.** Gesture the quantities.
- Put down the 1 + 2 = 3 card and say, **Here is another way to show 1 and 2.**
- Say, **This says 1 plus 2 equals 3.**
- Say, **See how the dots on each part match the numbers on the plus card. 1 plus 2 equals 3.** Gesture the quantities.
- Say, **Let's say this number sentence together. 1 plus 2 equals 3.** Repeat. Touch the numbers and the signs on the 1 + 2 = 3 card.

3 − 1 = 2

| 3 − 1 = 2 |
| Number Sentence Card |

- Say, **I can use this partner card to show a take-away problem, too. Watch what I do.**
- Say, **I have 3, but if I take away 1** (cover up the dot on the right), **I have 2 left.** Point to the remaining dot. Demonstrate again.
- Put down the 3 − 1 = 2 card and say, **Here is another way to show 3 take away 1. We call this minus.** Point to the minus sign. **What does this say?** Wait for a response.
- Say, **So this says 3 minus 1 equals 2.**
- Say, **Let's say this number sentence together. 3 minus 1 equals 2.** Repeat. Touch the numbers and the signs on the 3 − 1 = 2 card.

3 − 2 = 1

| 3 − 2 = 1 |
| Number Sentence Card |

- Say, **I can use this partner card to show another take-away problem. Watch what I do.**
- Say, **I have 3, but if I take away 2** (cover up the 2 dots on the right), **I have 1 left.** Point to the remaining dot. Demonstrate again.
- Put down the 3 − 2 = 1 card and say, **Here is another way to show 3 take away 2. This says 3 minus 2 equals 1.**
- Say, **Let's say this number sentence together. 3 minus 2 equals 1.** Repeat. Touch the numbers and the signs on the 3 − 2 = 1 card.

3 − 3 = 0

| 3 − 3 = 0 |
| Number Sentence Card |

- Put down the 3 − 3 = 0 card and say, **If I have 3 dots** (circle all 3 dots) **and take away 3 dots** (cover all 3 dots), **how many dots are left?** Wait for a response.

★ Review

- Put the 2 + 1 = 3, 1 + 2 = 3, 3 − 1 = 2, 3 − 2 = 1, and the 3 − 3 = 0 cards next to each other above the Partner Dot Card for 3.
- Say, **Let's say these one more time.** Gesture using the Partner Dot Card as you say each number sentence.

STORY PROBLEMS

MATERIALS: 3 of each: pencils, erasers, crayons, blocks; Number Sentence Cards

2 + 1

- Say, **Now I am going to tell you a story.** Give 2 pencils to a child sitting nicely and say, **_(Child)_ has 2 pencils. His/her teacher gave him/her 1 more pencil** (give the child another pencil). **How many pencils does _(child)_ have now?** Wait for a response.
- Hold up the 2 + 1 = 3 card and say, **That is right because 2 pencils plus 1 pencil equals 3 pencils.** Point to the numbers as you say them.

> 2 + 1 = 3
> Number Sentence Card

1 + 2

- Say, **Now I am going to tell you another story.** Give another child 1 eraser and say, **_(Child)_ has 1 eraser. His/her teacher gave him/her 2 more erasers** (give the child 2 more erasers). **How many erasers does _(child)_ have now?** Wait for a response.
- Hold up the 1 + 2 = 3 card and say, **That is right because 1 eraser plus 2 erasers equals 3 erasers.** Point to the numbers as you say them.

> 1 + 2 = 3
> Number Sentence Card

3 − 1

- Say, **This story is going to be a little different.** Give another child 3 crayons and say, **_(Child)_ has 3 crayons. The teacher takes 1 of his/her crayons** (take 1 crayon). **How many crayons does _(child)_ have now?** Wait for a response.
- Hold up the 3 − 1 = 2 card and say, **That is right because 3 crayons minus 1 crayon equals 2 crayons.**

> 3 − 1 = 2
> Number Sentence Card

3 − 2

- Say, **Here is 1 more story.** Give another child 3 blocks and say, **_(Child)_ has 3 blocks. The teacher takes 2 of his/her blocks** (take 2 blocks). **How many blocks does _(child)_ have now?** Wait for a response.
- Hold up the 3 − 2 = 1 card and say, **That is right because 3 blocks minus 2 blocks equals 1 block.**

> 3 − 2 = 1
> Number Sentence Card

NUMBER SENTENCES ON FINGERS

MATERIALS: Number Sentence Cards

2 + 1 = 3

- Hold up the 2 + 1 = 3 card. Say, **Watch me do this on my fingers.**
- Say, **The first number is 2. Here are 2 fingers.** Demonstrate on your right hand.
- Point to the + 1 on the card and say, **This says + 1, so I will put up 1 more finger.** Put up 1 finger on your left hand.
- Say, **How many fingers do I have up?** Wait for a response. **That is right because 2 plus 1 equals 3.** Emphasize 2 fingers, then 1 finger as you say, "2 plus 1" and both hands together as you say, "3."
- Say, **Now you try it. 2** (check all hands) **plus 1** (check all hands) **equals 3.** Repeat.
- *Note:* If children count 1, 2 as they put up the first addend, encourage them to be automatic—put 2 fingers up without counting.

> 2 + 1 = 3
> Number Sentence Card

1 + 2 = 3

- Hold up the 1 + 2 = 3 card. Say, **Watch me do this on my fingers.**
- Say, **The first number is 1. Here is 1 finger.** Demonstrate on your right hand.
- Point to the + 2 on the card and say, **This says + 2, so I will put up 2 more fingers.** Put up 2 fingers on your left hand.
- Say, **How many fingers do I have up?** Wait for a response. **That is right because 1 plus 2 equals 3.** Emphasize 1 finger, then 2 fingers as you say, "1 plus 2" and both hands together as you say, "3."
- Say, **Now you try it. 1** (check all hands) **plus 2** (check all hands) **equals 3.** Repeat.

> 1 + 2 = 3
> Number Sentence Card

3 − 1 = 2

- Put the 3 − 1 = 2 card in front of you. Say, **Watch me do this on my fingers.**
- Say, **The first number is 3. I will put 3 fingers on the table.** Demonstrate by putting 3 fingers down at once, not one at a time.
- Say, **Now I will take away 1 finger, like this!** Demonstrate holding 1 finger with your other hand, covering it.
- Say, **How many fingers do I have up?** Wait for a response. **That is right because 3 minus 1 equals 2.** Demonstrate again as you speak.
- Say, **Now you try it. 3** (check all hands) **minus 1** (check all hands) **equals 2.** Repeat.
- Repeat the previous steps for the 3 − 2 = 1 and 3 − 3 = 0 Number Sentence Cards.

> 3 − 1 = 2
> Number Sentence Card

WRITTEN NUMBERS

MATERIALS: Lesson 2 Activity Sheets, pencils, crayons

- Say, **Now we are going to practice writing numbers. Here is a paper with the numbers 2 and 3 on it. I would like you to trace over the numbers with your pencil. Do your best to stay on the dotted lines, but do not erase if you make a mistake. Just try it again! Under the dotted numbers is a place to copy the numbers on your own. Then write the number that tells us how many things are in the box.** Point to the pictures at the bottom of the page.
- As each child finishes, have him or her turn the paper over and say, **Write the missing number for these number sentences. If you do not know the missing number, then use your fingers to figure it out.** If some children finish early, then they can color the objects on the paper. Have children self-correct their papers by orally counting the objects. Do not let a child color an incorrect paper.

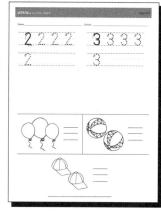
Lesson 2 Activity Sheet

Lesson 3

Learning Goals	Materials
Establish behavior boundaries	**COPY**
Count to 10 orally	Cardinality Chart
Build numbers 1-4 using blocks	Subitizing Circle Cards (#1-24)
Number recognition 0-4	Dot Chart for 4
Count and sequence to 4	Partner Dot Card for 4
Count to 4 on fingers	Lesson 3 Activity Sheet
Make numbers 1-4 on fingers	Number Sentence Cards: 3 + 1 = 4, 1 + 3 = 4, 4 − 1 = 3, 4 − 3 = 1, 4 − 4 = 0
Recognize quantities 1-4	
Partners of 4 (1/3)	**GATHER**
Story problems and number sentences: 4 family	White board magnetic easel
Perform number operations on fingers	10 interlocking blocks in plastic bag
Write the numerals 3, 4	4 of each: pencils, erasers, crayons, blocks
Connect the quantity to numerals 0-4	Pencils without erasers, crayons
Solve written number sentences: 2, 3, 4 (1/3) families	
	PREPARE
	Put student names on Lesson 3 Activity Sheets
	Number Recognition Cards (0-4); *See Chapter 1 for instructions on making Number Recognition Cards.*

ESTABLISHING BEHAVIOR BOUNDARIES

- Say, **When we work together, I expect you all to carefully listen and do your best. Okay? When I am talking, you should be listening. And when I call on one of you to speak, we will all listen to you. That is how we show respect for each other. It is important that we respect each other so we can all do our best.**

- To show me you are ready to learn, your body should be still, sitting just like this, feet under the table, bottom on the chair, and hands to myself. Wait for imitation. **Exactly! That shows me you are ready.**

COUNTING WARM-UP

- Say, **Let's count to 10 taking turns like last time. Each of you will say the next number when it is your turn. I will start ... 1.** Point to the child to your left. If a child says the incorrect number, then say, **That was a good try but the next number is __. Let's try again.** For each error, back up 2 children and repeat so the child has an opportunity to be successful. Take a turn yourself.

MAGIC NUMBER ACTIVITIES (Magic Number is 4)

★ Cardinality

MATERIALS: Cardinality Chart, 10 interlocking blocks in a plastic bag

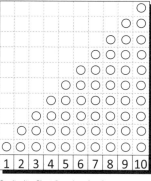
Cardinality Chart [number chart]

- Lay the Cardinality Chart on the table and say, **Here is our number chart. It has the numbers 1–10 across the bottom.** Run your finger across the numbers.
- Point to the number 1 and say, **The first number is 1. _(Child)_, please take out 1 block and put it on the chart.**
- Allow the child to take 1 block out of the plastic bag of interlocking blocks and put it on the chart.
- Point to the number 2 and say, **The next number is 2. _(Child)_, please take out 2 blocks and stick them together and put them on the chart.**
- Repeat with numbers 3 and 4.
- Say, **As we go up the number list on the number chart** (gesture up the number list on the Cardinality Chart), we add 1 block at a time. It is like climbing up steps, 1 at a time. Use gestures from Lesson 2 while saying, **See, 1 and 1 more is 2, 2 and 1 more is 3, 3 and 1 more is 4.** Touch the numeral 4 and run your finger up the 4 blocks.
- Say, **Let's try that again! 1 and 1 more is ___,** and so forth. If the children do not answer right away, then use previous gestures to scaffold the answer.

★ Sequencing and Number Recognition 0–4

MATERIALS: Number Recognition Cards 0–4, white board magnetic easel

Number Recognition Cards

- Say, **Say the numbers with me as I put them down. Do not go ahead of me. Ready? 1, 2, 3, 4!** Put the cards down horizontally as if building a number list. Orient the cards toward the children.
- Hold up the Number Recognition Card 4 and say, **What number is this?**
- Say, **That is right. This is the number 4. Our Magic Number today is 4.**
- Say, **This is how we write a 4. Watch carefully—down, across, and down.** Write a 4 on the white board magnetic easel.
- Say, **Now it is your turn! We are going to write a big number 4 in the air. Follow my finger. Ready? Copy me!**
- Trace a number 4 in the air backward (about 18" high) so that the children will see it in the correct orientation. Say, **Let's try that again!** Repeat twice.
- Pick up the cards, add the 0 card, and shuffle the cards for the number recognition game.
- Say, **Let's play a game. I will hold up a card with a number on it, and I want you to tell me the name of the number. I will point to you for your turn, okay? Everyone else say it in your mind, not out loud.**
- Say, **But if the Magic Number comes up, I want everyone to say the number, even if it is not your turn. Pay close attention so you will know when our Magic Number comes up!**
- Go around the group, showing each child a different number from 0 to 4 (not in order). Make sure all children can see the number. Go around the group three times.

★ Error Correction

- If the children do not all answer for the Magic Number, then say, **This is the Magic Number so everyone is supposed to answer. Let's try that again!** Hold up the number 4 and say, **What number is this?**

- If a child incorrectly answers, then say, **That is a good try but the number is ___.** Put the card down, hold it up again, and say, **What number is this?**

★ Finger Counting

- Say, **Let's count to 4 on our fingers. Watch me. 1, 2, 3, 4.** Demonstrate starting with index finger.
- Say, **Now you try! 1, 2, 3, 4.** Make sure all the children are correctly counting. **Let's try again. 1, 2, 3, 4.**

SUBITIZING QUANTITIES (1-4) ACTIVITIES

★ Finger Automaticity

MATERIALS: Number Recognition Cards 1-4

- Say, **Let's practice making numbers on our fingers quickly. Let's start with 1. This is how we make 1.** Put up an index finger. **Everyone show me 1 on your fingers.** Make sure everyone shows his or her index finger.
- Say, **This is how we make 2.** Show 2 on your fingers. **Everyone show me 2 on your fingers.** Correct any errors.
- Say, **This is how we make 3.** Show 3 on your fingers. **Everyone show me 3 on your fingers.** Correct any errors.
- Say, **This is how we make 4.** Show 4 on your fingers. **Everyone show me 4 on your fingers.** Correct any errors.
- Say, **Let's play our game. I want you to look at the number I show you and put up that many fingers. If I show you this card** (show the Number Recognition Card 4), **then how many fingers should you put up?** Wait for a response.

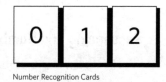
Number Recognition Cards

- Say, **That is right. Show me 4 fingers.** Make sure everyone shows 4 fingers.
- Shuffle the Number Recognition Cards and show them one at a time. Go through the cards three times, shuffling each time.
- Be sure that each child is holding up the correct number of fingers. If he or she is incorrect, then have him or her look at all the other children and correct him- or herself or let another child help him or her. Discourage counting and encourage automaticity.

★ Recognizing Sets

MATERIALS: Subitizing Circle Cards #1-24

Subitizing Circle Card

- Hold up Subitizing Circle Card # 19 and say, **Here is a card with 4 circles. See, here are 3 circles** (point to the bottom *triangle* of circles) **hiding inside the 4 circles. 3 and 1 more is 4.** Circle the dots as you say the quantities.
- Put the card down, hold it up, and say, **How many circles?** After 1 second, put the card down.
- Repeat with Subitizing Circle Cards #20-24.
- Shuffle the new cards into the pile toward the front.
- Say, **Now I am going to hold up a card with some circles on it, and I want you to tell me how many circles are on the card. Try to tell me as fast as you can without counting. I will point to you for your turn, okay? Everyone else, say it in your mind.**
- Go through the Subitizing Circle Cards in random order. Go around the group four times. Use the protocol from Lesson 1.

Number Sense Interventions Lesson 3   33

PARTNERS OF 4 ACTIVITIES

MATERIALS: Pencil, Partner Dot Cards for 4, Dot Chart for 4, Number Sentence Cards

★ Partners of 4 (3 + 1) Using Dot Chart for 4

- Put down the Dot Chart with 4 dots. Orient it horizontally.
- Say, **See, here are 4 dots. 1, 2, 3, 4.** Touch the dots as you count. **I can break them into 2 parts with my pencil.** Make sure 3 dots are at the child's left (your right).
- Say, **3 dots plus 1 dot equals 4 dots.** Circle dots on either side of the pencil as you say, "3 plus 1." Circle all the dots when you say, "4."
- Put down the matching Partner Dot Card for 4, oriented like the Dot Chart for 4. Gesture to each set of dots as you say, **See, 4 has 3 and 1 hiding inside it.**
- Say, **3 plus 1 equals 4.** Point to the sets of dots as you say, "3 plus 1" and circle the 4 dots as you say, "equals 4."
- Say, **How much is 3 plus 1?** Continue to gesture to the two sides of the card and then circle all the dots. Repeat.

★ Partners of 4 (1 + 3)

- Say, **Let's turn this Partner Dot Card over.** Rotate the card 180 degrees.
- Say, **Now this Partner Dot Card shows 1 plus 3 equals 4. See, 4 has 1 and 3 hiding inside it.** Continue gesturing to the sets of dots.
- Say, **How much is 1 plus 3?** Continue to gesture to the two sides of the card and then circle all the dots. Repeat.

★ Partners and Number Sentences

3 + 1 = 4

- Rotate the Partner Dot Card so 3 dots are on the child's left. Put down the 3 + 1 = 4 card and say, **Here is another way to show 3 plus 1.**
- Say, **This says 3 plus 1 equals 4.**
- Say, **See how the dots on each part match the numbers on the plus card. 3 plus 1 equals 4.** Gesture quantities.
- Say, **Let's say this number sentence together. 3 plus 1 equals 4.** Repeat. Touch the numbers and the signs on 3 + 1 = 4 card.

1 + 3 = 4

- Rotate the Partner Dot Card 180 degrees and say, **Now our card shows 1 plus 3.** Gesture quantities.
- Put down the 1 + 3 = 4 card and say, **Here is another way to show 1 plus 3.**
- Say, **This says 1 plus 3 equals 4.**

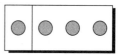

- Put down the Partner Dot Card and say, **See how the dots on each part match the numbers on the plus card. 1 plus 3 equals 4.** Gesture quantities.
- Say, **Let's say this number sentence together. 1 plus 3 equals 4.** Repeat. Touch the numbers and the signs on the 1 + 3 = 4 card.

4 − 1 = 3

- Say, **I can use this Partner Dot Card to show a take-away problem, too. Watch what I do.**
- Say, **I have 4, but if I take away 1** (cover up the dot on the left), **I have 3 left.** Point to the remaining dot. Demonstrate again.
- Put down the 4 − 1 = 3 card and say, **Here is another way to show 4 take away 1. This says 4 minus 1 equals 3.**
- Say, **Let's say this number sentence together. 4 minus 1 equals 3.** Repeat. Touch the numbers and the signs on the 4 − 1 = 3 card.

4 − 1 = 3
Number Sentence Card

4 − 3 = 1

- Say, **I can use this Partner Dot Card to show another take-away problem, too. Watch what I do.**
- Say, **I have 4, but if I take away 3** (cover up the 3 dots on the right), **I have 1 left.** Point to the remaining dot. Demonstrate again.
- Put down the 4 − 3 = 1 card and say, **Here is another way to show 4 take away 3. This says 4 minus 3 equals 1.**
- Say, **Let's say this number sentence together. 4 minus 3 equals 1.** Repeat. Touch the numbers and the signs on the 4 − 3 = 1 card.

4 − 3 = 1
Number Sentence Card

4 − 4 = 0

- Put down the Dot Chart for 4 and say, **If I have 4 dots** (circle all 4 dots) **and take away 4 dots** (cover all 4 dots), then **how many dots are left?** Wait for a response.
- Put down the 4 − 4 = 0 card and say, **So 4 minus 4 equals 0.** Repeat. Touch the numbers and the signs on the 4 − 4 = 0 card.

4 − 4 = 0
Number Sentence Card

Dot Chart for 4

★ Review

- Put the 3 + 1 = 4, 1 + 3 = 4, 4 − 1 = 3, 4 − 3 = 1, and 4 − 4 = 0 next to each other above the Partner Dot Card.
- Say, **Let's say these one more time.** Say each number sentence as written on the card and gesture using the Partner Dot Card as you say each number sentence.

STORY PROBLEMS

MATERIALS: 4 of each: pencils, erasers, crayons, blocks; Number Sentence Cards

3 + 1 = 4

- Say, **Now I am going to tell you a story.** Give a child sitting nicely 3 pencils and say, **_(Child)_ has 3 pencils. His/her teacher gave him/her 1 more pencil** (give the child another pencil). **How many pencils does _(child)_ have now?** Wait for a response.
- Hold up the 3 + 1 = 4 card and say, **That is right because 3 pencils plus 1 pencil equals 4 pencils.** Point to the numbers as you say them.

3 + 1 = 4
Number Sentence Card

1 + 3 = 4

- Say, **Now I am going to tell you another story.** Give another child 1 eraser and say, **_(Child)_ has 1 eraser. His/her teacher gave him/her 3 more erasers** (give the child 3 erasers). **How many erasers does _(child)_ have now?** Wait for a response.
- Hold up the 1 + 3 = 4 card and say, **That is right because 1 eraser plus 3 erasers equals 4 erasers.** Point to the numbers as you say them.

1 + 3 = 4
Number Sentence Card

Number Sense Interventions Lesson 3   35

4 − 1 = 3

- Say, **Here is another story.** Give another child 4 crayons and say, **_(Child)_ has 4 crayons. The teacher takes 1 of his/her crayons** (take 1 crayon). **How many crayons does _(child)_ have now?** Wait for a response.
- Hold up the 4 − 1 = 3 card and say, **That is right because 4 crayons minus 1 crayon equals 3 crayons.** Point to the numbers as you say them.

| 4 − 1 = 3 |
| Number Sentence Card |

4 − 3 = 1

- Say, **Here is another story.** Give another child 4 blocks and say, **_(Child)_ has 4 blocks. The teacher takes 3 of his/her blocks** (take 3 blocks). **How many blocks does _(child)_ have now?**
- Hold up the 4 − 3 = 1 card and say, **That is right because 4 blocks minus 3 blocks equals 1 block.**

| 4 − 3 = 1 |
| Number Sentence Card |

NUMBER SENTENCES ON FINGERS

MATERIALS: Number Sentence Cards, Partner Dot Cards for 4

3 + 1 = 4

- Hold up the 3 + 1 = 4 card. Say, **Watch me do this on my fingers.**
- Say, **The first number is 3. Here are 3 fingers.** Demonstrate on right hand.
- Say, **Now we will add 1 more finger.** Demonstrate on left hand.
- Say, **How many fingers do I have up?** Wait for a response. **That's right. 3 plus 1 equals 4.** Emphasize 3 fingers, then 1 finger as you say, "3 plus 1" and both hands together as you say, "4."
- Say, **Now you try it. Put up 3 fingers** (check all hands), **plus 1** (check all hands) **equals 4.**

| 3 + 1 = 4 |
| Number Sentence Card |

1 + 3 = 4

- Say, **We can see from our partner card that 3 plus 1 equals 4 and** (turn the Partner Dot Card 180 degrees) **1 plus 3 equals 4, so we show this either way—1 plus 3 or 3 plus 1.**
- Hold up the 1 + 3 = 4 card. Say, **Let's say this number sentence together. 1 plus 3 equals 4.**
- Say, **This time we will start by making 1 on this hand.** Demonstrate on right hand.
- Say, **Now we will add 3 more fingers, like this!** Demonstrate on left hand.
- Say, **How many fingers do I have up?** Wait for a response. **That is right. 1 plus 3 equals 4.** Emphasize 1 finger, then 3 fingers as you say, "1 plus 3" and both hands together as you say, "4."
- Say, **Now you try it. Put up 3 fingers** (check all hands), **plus 1** (check all hands), **equals 4.**

Partner Dot Card for 4

| 1 + 3 = 4 |
| Number Sentence Card |

4 − 1 = 3

- Put the 4 − 1 = 3 card in front of you. Say, **Watch me do this on my fingers.**
- Say, **The first number is 4. I will put 4 fingers on the table.** Demonstrate by putting 4 fingers down at once, not one at a time.
- Say, **Now we will take away 1 finger, like this!** Demonstrate holding 1 finger with your other hand, covering it.
- Say, **How many fingers do I have up?** Wait for a response. **That is right. 4 minus 1 equals 3.** Demonstrate again as you speak.
- Say, **Now you try it. 4** (check all hands) **minus 1** (check all hands) **equals 3.** Repeat.
- Repeat the previous steps for the 4 − 3 = 1 and 4 − 4 = 0 Number Sentence Cards.

| 4 − 1 = 3 |
| Number Sentence Card |

WRITTEN NUMBERS

MATERIALS: Lesson 3 Activity Sheets, pencils, crayons

- Say, **Now we are going to practice writing numbers. Here is a paper with the numbers 3 and 4 on it. I would like you to trace over the numbers with your pencil. Do your best to stay on the dotted lines, but do not erase if you make a mistake. Just try it again! Under the dotted numbers is a place to copy the numbers on your own. Then write the number that tells us how many things are in the box.** Point to the pictures at the bottom of the page.

- As each child finishes, have him or her turn the paper over and say, **Write the missing number for these number sentences. If you do not know the missing number, then use your fingers to figure it out.** If some children finish early, then they can color the objects on the paper. Have children self-correct their papers by orally counting the objects. Do not let a child color an incorrect paper.

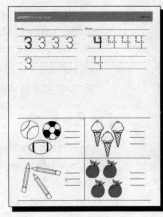

Lesson 3 Activity Sheet

Lesson 4

Learning Goals	Materials Needed
Establish behavior boundaries	**COPY**
Count to 20 orally	Hundreds Chart
Build numbers 1–5 using blocks	Cardinality Chart
Number recognition 0–5	Subitizing Circle Cards (#1–24)
Count and sequence to 5	Partner Dot Cards for 4
Before and after (1–5)	Dot Chart for 4
Count to 5 on fingers	Lesson 4 Activity Sheet
Make numbers 1–5 on fingers	Number Sentence Cards: 2 + 2 = 4, 4 − 2 = 2
Recognize quantities 0–4	**GATHER**
Partners of 4 (2/2)	White board magnetic easel
Story problems and number sentences: 4 family	15 interlocking blocks in a plastic bag
Perform number operations on fingers	4 crayons
Write the numerals 4 and 5	Pencils without erasers and crayons
Connect quantity to numerals 0–5	**PREPARE**
Solve written number sentences: 4 family	Put student names on Lesson 4 Activity Sheets.
	Number Recognition Cards (0–5); *See Chapter 1 for instructions on making Number Recognition Cards.*

ESTABLISHING BEHAVIOR BOUNDARIES

- Say, **Show me you are ready to learn.** Wait for proper posture. **Exactly! That shows me you are ready!**

Number Sense Interventions Lesson 4

COUNTING WARM-UP

MATERIALS: Hundreds Chart, white board magnetic easel

- Put up the Hundreds Chart to use for reference when needed.
- Say, **Let's count to 20 this time, taking turns. I will start. 1** (point to the child to your left). If a child says the incorrect number, then say, **That was a good try but the next number is ___. Let's try again.** For each error, back up 2 children and repeat so the child has an opportunity to be successful. Take a turn yourself.

1	2	3	4	5	6	7	8	9	10
11	12	13	14	15	16	17	18	19	20
21	22	23	24	25	26	27	28	29	30
31	32	33	34	35	36	37	38	39	40
41	42	43	44	45	46	47	48	49	50
51	52	53	54	55	56	57	58	59	60
61	62	63	64	65	66	67	68	69	70
71	72	73	74	75	76	77	78	79	80
81	82	83	84	85	86	87	88	89	90
91	92	93	94	95	96	97	98	99	100

Hundreds Chart

MAGIC NUMBER ACTIVITIES (Magic Number is 5)

★ Cardinality

MATERIALS: Cardinality Chart, 15 interlocking blocks in a plastic bag

- Lay the Cardinality Chart on the table and say, **Here is our number chart. It has the numbers 1-10 across the bottom.** Run your finger across the numbers.
- Point to the number 1 and say, **I will put 1 block on the chart and you can do the rest!**
- Point to the number 2 and say, **The next number is 2. (Child), please take out 2 blocks and put them on the chart.**
- Repeat the previous steps for the numbers 3, 4, and 5.
- Say, **As we go up the number list** (gesture up the Cardinality Chart), **we add 1 block at a time. It is like climbing up steps, 1 at a time.** Use gestures from Lesson 2 while saying, **See, 1 and 1 more is 2, 2 and 1 more is 3, and so forth to 5.** Touch the numeral 5 and run your finger up the 5 blocks.
- Say, **Let's try that again! 1 and 1 more is ___.** If the children do not answer right away, then use the previous gestures to scaffold the answer.

Cardinality Chart [number chart]

★ Sequencing and Number Recognition 0-5

MATERIALS: Number Recognition Cards 0-5, white board magnetic easel

- Say, **Say the numbers with me as I put them down. Do not go ahead of me. Ready? 1, 2, 3, 4, 5!** Put the cards down horizontally as if building a number list. Orient the cards toward the children.
- Hold up the Number Recognition Card 5 and say, **What number is this?** Wait for a response.
- Say, **That is right. This is the number 5. Our Magic Number today is 5.**
- Say, **This is how we write a 5. Watch carefully, down around and across.** Write a 5 on the white board magnetic easel.
- Say, **Now it is your turn! We are going to write a big number 5 in the air. Follow my finger. Ready? Copy me!**
- Trace a number 5 in the air backward (about 18" high) so that the children will see it in the correct orientation. Say, **Let's try that again!** Repeat twice.

Number Recognition Cards

- Pick up the cards, add the 0 card, and shuffle the cards for the number recognition game.
- Say, **Let's play our game! Remember, if the Magic Number comes up, then I want everyone to say the number, even if it is not your turn. What is the Magic Number for today?** Wait for a response. **Pay close attention so you will know when our Magic Number comes up!**
- If they do not all answer for the Magic Number, then say, **This is the Magic Number so everyone is supposed to answer.** Wait for a response. **Good! Let's try that again!** Hold up the number 5 and say, **What number is this?**
- Go around the group, showing each child a different number from 0 to 5 (not in order). Make sure all children can see the number. Go around the group three times.

★ Before and After

MATERIALS: Cardinality Chart, Number Recognition Cards 1–5

- Make sure each child gets a turn putting down a card. Use the Cardinality Chart for error correction.
- Put the Number Recognition Cards down in a row, in random order, and say, **Here are the numbers 1–5, but they are not in order.**
- Say, **Which number is the smallest amount?** Wait for a response. If a child says something other than "1," then refer to the Cardinality Chart and say, **Which number has the smallest number of circles?**
- Say, **That is right. 1 is the smallest number. (Child) , can you find the number 1 and put it right here?** Begin a number list.
- Point to the space right after 1 and say, **What number comes right after 1?** Wait for a response.
- Say, **That is right. 2 comes right after 1. (Child) , can you find the number 2 and put it right here?**
- Continue until all numbers are down in order.
- Pick up the cards and shuffle them. Say, **Now we are going to play the before game.**
- Put down the cards in random order and say, **Which number here is the biggest amount?** Wait for a response.
- Say, **That is right. 5 is the biggest number. (Child) , can you find the number 5 and put it right here?** Begin a number list.
- Point to the space right before 5 and say, **What number comes right before 5?** Wait for a response. Say, **That is right. 4 comes right before 5. (Child) , can you find the number 4 and put it right here?**
- Continue until all numbers are down in order.
- Say, **Let's count down to check: 5, 4, 3, 2, 1. Does that count right?** Wait for a response.

Number Recognition Cards

★ Finger Counting

- Say, **Let's count to 5 on our fingers. Watch me. 1, 2, 3, 4, 5.** Demonstrate starting with the index finger.
- Say, **Now you try! 1, 2, 3, 4, 5. Let's try again. 1, 2, 3, 4, 5.** Make sure all the children are correctly counting.
- Say, **How many fingers do I have on this hand?** Wait for a response.
- Say, **That is right! I have 5 fingers on this hand.**
- Hold up your left hand and say, **How about on this hand? Let's count again! 1, 2, 3, 4, 5.** Demonstrate starting with the index finger. **How many fingers do I have on this hand?** Wait for a response.
- Say, **That is right! I have 5 fingers on each hand.** Hold up each hand one at a time. **Do you have 5 fingers on each hand?** Wait for a response.

- When you finish counting say, **From now on, we do not need to count to find out how many fingers are on one hand. We know! How many fingers are on one hand?** Wait for a response.
- Hold up one hand and say, **How many fingers on this hand?** Wait for a response. Put that hand down and hold up the other hand and say, **How many fingers on this hand?** Wait for a response.
- Say, **Suppose I hold my hand behind my back. How many fingers are on my hand?** Wait for a response. **Suppose I put it under my arm. How many fingers are on my hand?** Wait for a response.

SUBITIZING QUANTITIES (1-5) ACTIVITIES

★ Finger Automaticity

MATERIALS: Number Recognition Cards 1-5

- Say, **Let's practice making numbers on our fingers quickly.**
- Say, **Show me 1 on your fingers.** Make sure everyone shows his or her index finger.
- Repeat through 5 fingers.
- Say, **Let's play our game. I want you to look at the number I show you and you put up that many fingers.**
- Shuffle the Number Recognition Cards and show them one at a time. Go through the cards three times, shuffling each time.
- Be sure that each child is holding up the correct number of fingers. If he or she is incorrect, then have him or her look at all the other children and correct him- or herself or let another child help. Discourage counting and encourage automaticity.

★ Recognizing Sets

MATERIALS: Subitizing Circle Cards #1-24

- Show the Subitizing Circle Card #21 (with 4 circles) and say, **Here is a card with 4 circles. See, here are 3 circles hiding inside the 4 circles. 3 and 1 more is 4.** Circle the dots as you say the quantities.
- Put the card down, hold it up, and say, **How many?** Wait for a response. Repeat with Subitizing Circle Cards with 4 circles.
- Show card #21 again and say, **Let's look at this one again. See, here are 2 circles** (circle the circles in the left column) **and 2 circles** (circle the circles in the right column) **hiding inside the 4 circles. 2 and 2 is 4.** Circle the dots as you say the quantities.
- Put the card down, hold it up, and say, **How many?** Wait for a response.
- Repeat with Subitizing Circle Cards #19-24.
- Shuffle the new cards into the front of the pile.
- Say, **Now I am going to hold up a card with some circles on it, and I want you to tell me how many circles are on the card. Try to tell me as fast as you can without counting. I will point to you for your turn, okay? Everyone else, say it in your mind.**
- Go through the Subitizing Circle Cards in random order. Go around the group four times. Use protocol from Lesson 1.

PARTNERS OF 4 ACTIVITIES

MATERIALS: Pencil, Partner Dot Card for 4, Dot Chart for 4, Number Sentence Cards

★ Partners of 4 (2 + 2) Using Dot Chart for 4

Dot Chart for 4

- Say, **Let's put the pencil in the middle of the dots.** Put the pencil between the 2 middle dots on the Dot Chart for 4.
- Point to the 2-dot side and say, **How many dots are on this part?** Point to the other 2-dot side and say, **How many dots are on this part?**
- Say, **How many dots are there in all?** Wait for a response. Say, **That is right! 2 dots plus 2 dots equals 4 dots.** Circle dots on either side of the pencil as you say, "2 plus 2." Circle all the dots when you say, "4."

Partner Dot Card for 4

- Put down the Partner Dot Card for 4 and say, **This partner card shows 2 plus 2 equals 4. See, 4 has 2 and 2 hiding inside it.** Point to each set of dots, 2 then 2 while saying, "2 plus 2" and circle the 4 dots as you say, "equals 4."
- Say, **How much is 2 plus 2?** Continue to gesture to the 2 sides of the card and then circle all the dots. Repeat.

★ Partners and Number Sentences

2 + 2 = 4

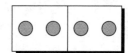
Partner Dot Card for 4

- Put down the 2 + 2 = 4 card and say, **Here is another way to show 2 and 2.**
- Say, **This says 2 plus 2 equals 4.**
- Put down the Partner Dot Card and say, **See how the dots on each part match the numbers on the number sentence? 2 plus 2 equals 4.** Gesture the quantities.

2 + 2 = 4
Number Sentence Card

- Say, **Let's say this number sentence together. 2 plus 2 equals 4.** Repeat. Touch the numbers and the signs on the 2 + 2 = 4 card.

4 − 2 = 2

- Say, **I can use this partner card to show a take-away problem, too. Watch what I do.**
- Say, **I have 4, but if I take away 2** (cover up the dots on the right), **then I have 2 left.** Point to the remaining dots. Demonstrate again.

4 − 2 = 2
Number Sentence Card

- Put down the 4 − 2 = 2 card and say, **Here is another way to show 4 take away 2. This says 4 minus 2 equals 2.**
- Say, **Let's say this number sentence together. 4 minus 2 equals 2.** Repeat. Touch the numbers and the signs on the 4 − 2 = 2 card.

★ Review

- Put the 2 + 2 = 4 and the 4 − 2 = 2 cards next to each other above the Partner Card Dot.
- Say, **Let's say these one more time.** Say each number sentence and gesture using the Partner Dot Card as you say each number sentence.

STORY PROBLEMS

MATERIALS: 4 crayons, Number Sentence Cards

2 + 2 = 4

- Say, **Now I am going to tell you another story.** Give 2 children 1 crayon each and say, **Only 2 children have crayons. The teacher gave 2 more children crayons.** Give the other 2 children crayons. **How many children now have crayons?** Wait for a response.
- Hold up the 2 + 2 = 4 card and say, **That is right because 2 plus 2 equals 4.** Point to the numbers as you say them. Let them keep the crayons for the next problem.

4 − 2 = 2

- Say, **This story is going to be a little different. 4 children have their hands folded. 2 children unfold their hands.** Touch the hands of 2 children and motion to unfold them. **How many children have their hands folded now?** Wait for a response.
- Hold up the 4 − 2 = 2 card and say, **That is right because 4 children minus 2 children equals 2 children.** Gesture to the children who participated.

NUMBER SENTENCES ON FINGERS

MATERIALS: Number Sentence Cards

2 + 2 = 4

- Hold up the 2 + 2 = 4 card. Say, **Watch me do this on my fingers.**
- Say, **The first number is 2. Here are 2 fingers.** Demonstrate on your right hand.
- Say, **Now we will add 2 more fingers.** Demonstrate by showing 2 fingers on your left hand.
- Say, **How many fingers do I have up?** Wait for a response. **That is right. 2 plus 2 equals 4.** Emphasize 2 fingers, then 2 fingers as you say, "2 plus 2" and both hands as you say, "4."
- Say, **Now you try it. Put up 2 fingers** (check all hands) **plus 2 more** (check all hands) **equals 4.**

WRITTEN NUMBERS

MATERIALS: Lesson 4 Activity Sheets, pencils, crayons

- Say, **Now we are going to practice writing numbers. Here is a paper with the numbers 4 and 5 on it. I would like you to trace over the numbers with your pencil. Do your best to stay on the dotted lines, but do not erase if you make a mistake. Just try it again! Under the dotted numbers is a place to copy the numbers on your own. Then write the number that tells us how many things are in the box.** Point to the pictures at the bottom of the page.
- As each child finishes, have him or her turn the paper over and say, **Write the missing number for these number sentences. If you do not know the missing number, then use your fingers to figure it out.**

Lesson 4 Activity Sheet

Lesson 5

Learning Goals

Establish behavior boundaries
Count to 20 orally
Number recognition 0–5
Count and sequence to 5
Before and after (1–5)
Count to 5 on fingers
Make numbers 1–5 on fingers
Recognize quantities 0–5
Partners of 5 (4/1, 3/2) using Five Frame
Perform number operations on fingers
Write the numerals 0–5
Connect quantity to numerals 0–5
Solve written number sentences: 5 family (plus)

Materials

COPY
Cardinality Chart
Hundreds Chart
Subitizing Circle Cards (#1–28)
Lesson 5 Activity Sheet
Number Sentence Cards: 1 + 4 = 5, 2 + 3 = 5, 3 + 2 = 5, 4 + 1 = 5

GATHER
White board magnetic easel
Pencils without erasers and crayons

PREPARE
Put student names on Lesson 5 Activity Sheets.
Number Recognition Cards (0–5); *See Chapter 1 for instructions on making Number Recognition Cards.*
Teacher Five Frame Mat with 5 two-colored dots; *See Chapter 1 for instructions on making the Five Frame Mat and two-colored dots.*
Student Five Frame Trays with 5 two-colored magnetic dots (1 set per student)

ESTABLISHING BEHAVIOR BOUNDARIES

- Say, **Show me you are ready to learn.** Wait for proper posture. **Exactly! That shows me you are ready!**

COUNTING WARM-UP

MATERIALS: Hundreds Chart, white board magnetic easel

- Put up the Hundreds Chart to use for reference when needed.
- Say, **Let's count to 20 this time, taking turns. I will start. 1** (point to the child to your left). If a child says the incorrect number, then say, **That was a good try but the next number is ___. Let's try again.** For each error, back up 2 children and repeat so the child has an opportunity to be successful. Take a turn yourself.

Hundreds Chart

MAGIC NUMBER ACTIVITIES (Magic Number is 5)

- *Note:* A Cardinality Chart activity will not be completed in this lesson.

★ Sequencing and Number Recognition 0–5

MATERIALS: Number Recognition Cards 0–5, white board magnetic easel

- Say, **Say the numbers with me as I put them down. Do not go ahead of me. Ready? 1, 2, 3, 4, 5!** Put the cards down horizontally as if building a number list. Orient the cards toward the children.

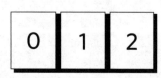

Number Recognition Cards

Number Sense Interventions Lesson 5   43

- Hold up the Number Recognition Card 5 and say, **What number is this?** Wait for a response.
- Say, **That is right. This is the number 5. Our Magic Number today is 5.**
- Say, **This is how we write a 5. Watch carefully, down around and across.** Write a 5 on the white board magnetic easel.
- Say, **Now it is your turn! We are going to write a big number 5 in the air. Follow my finger. Ready? Copy me!**
- Trace a number 5 in the air backward (about 18" high) so that the children will see it in the correct orientation. Say, **Let's try that again!** Repeat twice.
- Pick up the cards, add the 0 card, and shuffle the cards for the number recognition game.
- Say, **Let's play our game! Remember, if the Magic Number comes up, then I want everyone to say the number, even if it is not your turn. What is the Magic Number for today?** Wait for a response. **Pay close attention so you will know when our Magic Number comes up!**
- If they do not all answer for the Magic Number, then say, **This is the Magic Number so who is supposed to answer?** Wait for a response. **Good! Let's try that again!** Hold up the number 5 and say, **What number is this?**
- Go around the group, showing each child a different number from 0 to 5 (not in order). Make sure all children can see the number. Go around the group three times.

★ Before and After (Mixed)

- Make sure each child gets a turn putting down a Number Recognition Card. Use the Cardinality Chart for error correction.
- Put the Number Recognition Cards down in a row in random order and say, **Here are the numbers 1–5, but they are not in order.**
- Pull out the number 3 and say, **What number comes right before 3?** Wait for a response. Point to the space before 3.
- Say, **That is right. 2 comes right before 3.** If someone answers "2," then let him or her take the 2 and put it before the 3. If no one answers "2," then say, **When I count, what number do I say right before I say, 3. 1, ____, 3.** Wait for a response.
- Say, **What number comes right after 3?** Wait for a response. **1, 2, 3, ____?**
- Say, **That is right. 4 comes right after 3.** Have a child who correctly answered place the card in the correct order.
- Say, **What number comes right before 2?** Wait for a response. Point to the space before 2.
- Say, **That is right. 1 comes right before 2.** Have a child who correctly answered place the card.
- Say, **What number comes right after 4?** Wait for a response. **3, 4, ____?**
- Say, **That is right. 5 comes right after 4.** Have a child who correctly answered place the card in the correct order.
- Say, **Let's say our numbers to check. 1, 2, 3, 4, 5. Does that count right?** Wait for a response.

SUBITIZING QUANTITIES (1–5) ACTIVITIES

★ Finger Automaticity

MATERIALS: Number Recognition Cards

- Say, **Let's play our finger game. I want you to look at the number I show you and put up that many fingers.**

Number Recognition Cards

- Shuffle the Number Recognition Cards and show them one at a time. Go through the cards three times, shuffling each time.
- Be sure that each child is holding up the correct number of fingers. If he or she is incorrect, then have him or her look at all the other children and correct him- or herself or let another child help him or her. Discourage counting and encourage automaticity.

★ Recognizing Sets

Subitizing Circle Card

MATERIALS: Subitizing Circle Cards #1-28

- Show Subitizing Circle Card #28 and say, **Here is a card with 5 circles. See, here are 4 circles** (at the bottom) **and 1 circle hiding inside the 5 circles. 4 and 1 more is 5.** Circle the dots as you say the quantities.
- Put the card down, hold it up, and say, **How many?** Wait for a response.
- Repeat with cards #26 and #27.
- Show card #25 and say, **Here is another card with 5 circles. See, here are 3 and 2 circles hiding inside the 5 circles. 3 and 2 more is 5.** Circle the dots as you say the quantities.
- Put the card down, hold it up, and say, **How many?** Wait for a response.
- Say, **Now I am going to hold up a card with some circles on it, and I want you to tell me how many circles are on the card. Try to tell me as fast as you can without counting. I will point to you for your turn, okay? Everyone else, say it in your mind.**
- Go through all the Subitizing Circle Cards in random order. Go around the group four times. Use protocol from Lesson 1.

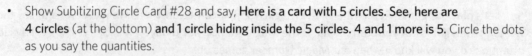

COMPARING QUANTITIES—MORE

MATERIALS: Subitizing Circle Cards #1-28

Subitizing Circle Card Subitizing Circle Card

- Hold up a 1-circle card and a 5-circle card and say, **Point to the card that has more circles—this or this.** Push one card, then the other card forward when you say, "this or this."
- Say, **That is right! 5 is more than 1.**
- Hold up 2 cards with 2 circles and say, **Point to the card that has more circles, this or this.** Wait for a response. If no one uses the word *equal,* then say, **When 2 cards have the same number of circles, we say they are equal.**
- Say, **That is right! 2 equals 2.**
- Continue through the Subitizing Circle Cards, taking pairs of cards as they come up saying, **Which has more?** and responding with, **That is right! ____ is more than ____.** Some children may naturally start saying the numbers but do not require it.
- Show 5 pairs after the first 2 models.

PARTNERS OF 5 ACTIVITIES

MATERIALS: Teacher Five Frame Mat with 5 two-colored dots, white board magnetic easel, student Five Frame Trays with 5 two-colored dots, Number Sentence Cards

Number Sense Interventions Lesson 5   45

★ Introduction to Five Frames

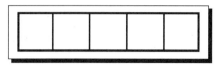

Five Frame Mat

- Put down the Five Frame Mat on the table. Say, **Today we are going to put our dots on this frame.** Point to the row and say, **See this row. It has 5 spaces—just like your hand has 5 fingers. 1, 2, 3, 4, 5.** Touch each space with each of your fingers on your right hand as you count pinky to thumb, left to right.

- Say, **How many spaces are in this row?** Wait for a response. **When we see this row all filled in, we do not need to count it because we know it is 5, just like our hand has 5 fingers!**

- Say, **We will use this frame to make partners of 5. I am going to put it on my board so everyone can see!** Put the Five Frame Mat up on the white board magnetic easel.

- Say, **Here are 5 red dots. I am going to turn 1 over so it is yellow.** Turn over the first dot. **Now I have 1 yellow dot and 4 red dots. 1 plus 4 equals 5.** Gesture to the dots.

- Put down the 1 + 4 = 5 card and say, **Here is the number sentence that says 1 plus 4 equals 5.** Gesture to the numbers on the card.

  $1 + 4 = 5$
  Number Sentence Card

- Say, **I am going to give you each a frame and 5 dots.** Pass out the Five Frame Trays with 5 magnetic dots all on the red side.

- Say, **Here is another number sentence that equals 5.** Put down the 2 + 3 = 5 card. **2 plus 3 equals 5. Let's make the partners on your frame. I will make it on my frame, too.**

- Turn over 2 dots to the yellow side. Check to see that all the children have done the same.

- Say, **Great job! 2 plus 3 equals 5.** Circle the quantities as you say them.

- Say, **Let's try another one!** Repeat with 3 + 2 = 5 and 4 + 1 = 5 cards.

NUMBER SENTENCES ON FINGERS

MATERIALS: Number Sentence Cards

Note: Encourage automaticity. This activity is different as both addends are shown on one hand.

- Hold up one hand and say, **How many fingers are on one hand?** Wait for a response.
- Say, **That is right! Just like the 5 dots on my frame. So, I can show the partners for 5 on my hand.**

4 + 1 = 5

- Hold up the 4 + 1 = 5 card. Say, **Watch me do this on my fingers.**
- Say, **The first number is 4. Here are 4 fingers.** Demonstrate on right hand.
- Say, **Now we will add 1 more finger.** Demonstrate by extending thumb on right hand.
- Say, **How many fingers do I have up?** Wait for a response. **That is right. 4 plus 1 equals 5.** Emphasize 4 fingers, then thumb as you say, "4 plus 1" and entire hand as you say, "5."
- Say, **Now you try it. Say it with me as you show it.** Repeat with children copying. Repeat again.

  $4 + 1 = 5$
  Number Sentence Card

1 + 4 = 5

- Hold up the 1 + 4 = 5 card. Say, **We can show this number sentence with the same fingers.**
- Say, **1** (wiggle thumb) **plus 4** (wiggle 4 fingers) **equals 5** (extend 5 fingers). Repeat.
- Say, **Now you try it. Say it with me as you show it.** Repeat with children copying. Repeat again.

  $1 + 4 = 5$
  Number Sentence Card

3 + 2 = 5

- Hold up the 3 + 2 = 5 card. Say, **Watch me do this on my fingers.** Touch index finger to thumb.
- Say, **The first number is 3. Here are 3 fingers** (wiggle 3 free fingers).
- Say, **Plus 2** (open and close thumb and index finger) **equals 5** (extend all 5 fingers).
- Say, **How many fingers do I have up?** Wait for a response. **That is right. 3 plus 2 equals 5.** Gesture as you say the number sentence.
- Say, **Now you try it. Say it with me as you show it.** Repeat with children copying. Repeat again.

3 + 2 = 5
Number Sentence Card

2 + 3 = 5

- Hold up the 2 + 3 = 5 card. Say, **We can show this number sentence with the same fingers.**
- Say, **2** (open and close thumb and index finger) **plus 3** (wiggle 3 fingers) **equals 5** (extend 5 fingers). Repeat.
- Say, **Now you try it. Say it with me as you show it.** Repeat with children copying. Repeat again.

2 + 3 = 5
Number Sentence Card

WRITTEN NUMBERS

MATERIALS: Lesson 5 Activity Sheets, pencils, crayons

- Say, **Now we are going to practice writing numbers. Here is a paper with all our Magic Numbers. I would like you to trace over the numbers with your pencil. Do your best to stay on the dotted lines, but do not erase if you make a mistake. Just try it again! Under the dotted numbers is a place to copy the numbers on your own.**

- As each child finishes, have him or her turn the paper over and say, **Write the missing number for these number sentences. If you do not know the missing number, then use your fingers to figure it out.**

Lesson 5 Activity Sheet

Lesson 6

Learning Goals	Materials
Establish behavior boundaries	**COPY**
Count to 20 orally	Hundreds Chart
Build numbers 1–6	Cardinality Chart
Number recognition 0–6	Subitizing Circle Cards (#1–28)
Count and sequence to 6	Lesson 6 Activity Sheet
Before and after (1–6)	Number Sentence Cards: 1 + 4 = 5, 2 + 3 = 5, 3 + 2 = 5, 4 + 1 = 5
Count to 6 on fingers	**GATHER**
Make numbers 1–6 on fingers	White board magnetic easel
Recognize quantities 1–6	Farm scene with 5 pig magnets
Story problems and number sentences: 5 family (plus)	21 interlocking blocks (20 of same color, 1 of different color) in a plastic bag
Perform number operations on fingers	Pencils without erasers and crayons
Write the numerals 5 and 6	**PREPARE**
Connect quantity to numerals 0–6	Put student names on Lesson 6 Activity Sheets.
Solve written number sentences: 5 family (plus)	Number Recognition Cards (0–6); *See Chapter 1 for instructions on making Number Recognition Cards.*
	Teacher Five Frame Mat with 5 two-colored magnetic dots; *See Chapter 1 for instructions on making the Five Frame Mat and two-colored dots.*
	Student Five Frame Trays with 5 two-colored magnetic dots

ESTABLISHING BEHAVIOR BOUNDARIES

- Say, **Show me you are ready to learn.** Wait for proper posture. **Exactly! That shows me you are ready!**

COUNTING WARM-UP

MATERIALS: Hundreds Chart, white board magnetic easel

- Put up the Hundreds Chart to use for reference when needed.
- Say, **Let's count to 20 this time, taking turns. I will start. 1** (point to the child to your left). If a child says the incorrect number, then say, **That was a good try but the next number is ____. Let's try again.** For each error, back up 2 children and repeat so the child has an opportunity to be successful. Take a turn yourself.

Hundreds Chart

MAGIC NUMBER ACTIVITIES (Magic Number is 6)

★ Cardinality

MATERIALS: Cardinality Chart, 21 interlocking blocks (20 of same color, 1 of different color) in a plastic bag

- Lay the Cardinality Chart on the table and say, **Here is our number chart. It has the numbers 1-10 across the bottom.** Run your finger across the numbers.
- Say, **I would like you each to make a stick of blocks and put them down on the table in front of you.   (Child 1) , make a stick of 2.   (Child 2) , make a stick of 3.   (Child 3) , make a stick of 4.   (Child 4) , make a stick of 5.**
- Make a stick of 5 while they are building.
- When all children are completed, point to the numeral 1 on the Cardinality Chart and say, **The number 1 has 1 block.** Place 1 block. **The number 2 has two blocks.   (Child 1) , put down your 2 blocks.** Continue until all blocks are placed.
- Point to the 6 and say, **This is the number 6. 6 is 5 and 1 more.**
- Hold up the stick of 5 and say, **Here are 5 blocks. 5 and 1 more** (put on the different colored block) **is 6.**
- Put the stick of 6 on the chart.
- Say, **As we go up the number list** (gesture up the Cardinality Chart), **we add 1 block at a time. It is like climbing up steps, 1 at a time.** Use gestures from Lesson 2 while saying, **See, 1 and 1 more is 2, 2 and 1 more is 3, and so forth to 6.** Touch the numeral 6 and run your finger up the 6 blocks.
- Say, **Let's try that again! 0 and 1 more is _____.** If they do not answer right away, then use the previous gestures to scaffold the answer.

Cardinality Chart [number chart]

★ Sequencing and Number Recognition 0-6

MATERIALS: Number Recognition Cards 0-6

- Say, **Say the numbers with me as I put them down. Do not go ahead of me. Ready? 1, 2, 3, 4, 5, 6!** Put the cards down horizontally as if building a number list. Orient the cards toward the children.
- Hold up the Number Recognition Card 6 and say, **What number is this?** Wait for a response.
- Say, **That is right. This is the number 6. Our Magic Number today is 6.**
- Say, **Let's play our game!** Follow error correction guidelines.
- Go around the group, showing each child a different number from 0 to 6 (not in order). Make sure all children can see the number. Go around the group three times.

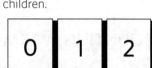
Number Recognition Cards

★ Before and After (Mixed)

MATERIALS: Number Recognition Cards 1-6

Note: Make sure each child gets a turn putting down a Number Recognition Card. Use the Cardinality Chart for error correction.

- Put the Number Recognition Cards down in a row in random order and say, **Here are the numbers 1-6, but they are not in order.**
- Pull out the number 4 and say, **What number comes right before 4?** Wait for a response. Point to the space before 4.

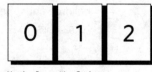
Number Recognition Cards

Number Sense Interventions Lesson 6

- Say, **That is right. 3 comes right before 4.** If someone answers, "3," then let him or her take the 3 and put it before the 4. If no one answers, "3," then say, **When I count, what number do I say right before I say 4? 1, 2, ___, 4.** Wait for a response.
- Say, **What number comes right after 4?** Wait for a response. **1, 2, 3, 4 ___?**
- Say, **That is right. 5 comes right after 4.** Have a child who correctly answered place the card in the correct order.
- Say, **What number comes right before 3?** Wait for a response. Point to the space before 3.
- Say, **That is right. 2 comes right before 3.** Have a child who correctly answered place the card in the correct order.
- Say, **What number comes right after 5? 3, 4, 5, ___.** Wait for a response.
- Say, **That is right. 6 comes right after 5.** Have a child who correctly answered place the card in the correct order.
- Continue with what comes before 2.
- Say, **Let's say our numbers to check. 1, 2, 3, 4, 5, 6. Does that count right?** Wait for a response.

★ Finger Counting

- Say, **Let's count to 6 on our fingers. Watch me first. 1, 2, 3, 4, 5. I have used up all my fingers! There are just 5 fingers on one hand. I need to use my other hand to show 6 fingers.**
- Say, **Watch this!** Hold up 5 fingers and say, **5.** Hold up the index finger of your other hand and say, **6. Now you try. 5, 6.** Make sure all the children hold up one hand when they say, "5" and add the index finger of the other hand when they say, "6."
- Say, **We do not need to count the 5 fingers on this hand because we know there are 5. So, we can just say 5 when we hold up all the fingers on one hand.**
- Say, **Let's try counting to 6 again, starting with 5. Ready? 5, 6.** Repeat.
- Say, **This is what 6 looks like on our fingers.** Demonstrate 6 all at once. **So if I say, "Show 6 on your fingers," what will you do?** Make sure every child shows 6 fingers as one hand and 1 finger.

SUBITIZING QUANTITIES (1–6) ACTIVITIES

MATERIALS: Number Recognition Cards 1–6

★ Finger Automaticity

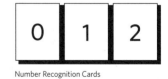

Number Recognition Cards

- Say, **Let's practice making numbers on our fingers quickly.**
- Say, **Now we know how to make 6 on our fingers. Let's make 6.** Demonstrate and correct any errors.
- Say, **Let's play our game. I want you to look at the number I show you and put up that many fingers.**
- Shuffle the Number Recognition Cards and show them one at a time. Go through the cards two times.
- Be sure that each child is holding up the correct number of fingers. If he or she is incorrect, then have him or her look at all the other children and correct him- or herself or let another child help him or her. Discourage counting and encourage automaticity.

50 Jordan and Dyson

★ Recognizing Sets

MATERIALS: Subitizing Circle Cards #1–28

Subitizing Circle Card

- Say, **Now I am going to hold up a card with some circles on it, and I want you to tell me how many circles are on the card. Try to tell me as fast as you can without counting. I will point to you for your turn, okay? Everyone else, say it in your mind.**
- Go through the Subitizing Circle Cards in random order. If a child gets a 4 or 5 card incorrect, then point out the smaller sets in the total.
- Go around the group four times.

COMPARING QUANTITIES—LESS

MATERIALS: Subitizing Circle Cards

Subitizing Circle Card   Subitizing Circle Card

- Hold up a 1-circle card and a 5-circle card and say, **Today we are going to do something different. Listen carefully to my words. Point to the card that has less circles—this or this?** Push one card, then the other card forward when you say, "this or this?"
- Say, **That is right! 1 is less than 5.**
- Hold up 2 cards with 2 circles and say, **Point to the card that has less circles—this or this?** If no one uses the word *equal*, then say, **When 2 cards have the same number of circles, we say they are equal.**
- Say, **That is right! 2 equals 2.**
- Continue through the Subitizing Circle Cards, taking pairs of cards as they come up saying, **Which has less?** and responding with, **That is right! ___ is less than ___.** Some children may naturally start saying the numbers but do not require it.
- Show 5 pairs after the first 2 models.

STORY PROBLEMS

MATERIALS: White board magnetic easel, farm scene with 5 pig magnets, Number Sentence Cards, Five Frame Trays with 5 two-colored magnetic dots

- Say, **Today we are going to tell stories about the farm!** Put the farm background with the pen on the white board magnetic easel.
- Model each story on the farm scene as you tell it.

2 + 3 = 5

2 + 3 = 5
Number Sentence Card

- Say, **Listen to this story. There are some pigs at the farm. 2 are in the pen and 3 are outside the pen. How many pigs are there altogether** (circle all the pigs as you say, "altogether")**?** Wait for a response. **That is right. 2 pigs plus 3 pigs equals 5 pigs altogether.**
- Put down the Number Sentence Cards: 1 + 4 = 5, 2 + 3 = 5, 3 + 2 = 5, and 4 + 1 = 5. Say, **Which number sentence goes with this story? 2 were in the pen and 3 were outside the pen. That is right. 2 plus 3 equals 5 altogether.** Point to the numbers and signs on the 2 + 3 = 5 card as you say them.
- Say, **Let's show that number sentence on our frames.** Help children who are struggling.

Number Sense Interventions Lesson 6

3 + 2 = 5

- Say, **Listen to this story. There are 3 pigs in the pen. The farmer brings 2 more into the pen. How many pigs are in the pen now?** Wait for a response. **That is right. 3 pigs plus 2 pigs equals 5 pigs.**
- Say, **Which number sentence goes with this story? There were 3 pigs in the pen and 2 more pigs were brought in.** Wait for a response. **That is right. 3 plus 2 equals 5.**
- Say, **Let's show that number sentence on our frames.** Help children who are struggling.

1 + 4 = 5

- Say, **Listen to this story. There is 1 pig on the road. 4 more pigs came to join him. How many pigs are on the road now?** Wait for a response. **That is right. 1 pig plus 4 pigs is 5 pigs.**
- Say, **Which number sentence goes with this story? There was 1 pig on the road and 4 more pigs came to join him.** Wait for a response. **That is right. 1 plus 4 equals 5.**
- Say, **Let's show that number sentence on our frames.** Help children who are struggling.

4 + 1 = 5

- Hold up the 4 + 1 = 5 card. Say, **Who would like to tell a story for this number sentence?**
- Choose a child to tell the story and guide him or her if he or she needs help. Model on the Five Frames Mat.

NUMBER SENTENCES ON FINGERS

MATERIALS: Number Sentence Cards

4 + 1 = 5

- Hold up the 4 + 1 = 5 card. Say, **Let's show this on our fingers. 4 plus 1 equals 5.** Emphasize 4 fingers, then thumb as you say, "4 plus 1" and the entire hand as you say, "5." Repeat twice.

1 + 4 = 5

- Hold up the 1 + 4 = 5 card. Say, **Let's show this on our fingers. 1 plus 4 equals 5.** Emphasize thumb, then fingers as you say, "1 plus 4" and the entire hand as you say, "5." Repeat twice.

3 + 2 = 5

- Hold up the 3 + 2 = 5 card. Say, **Let's show this on our fingers. 3 plus 2 equals 5.** Touch index finger to thumb and wiggle 3 fingers, open and close thumb and index finger, extend all 5 fingers. Repeat twice.

2 + 3 = 5

- Hold up the 2 + 3 = 5 card. Say, **Let's show this on our fingers. 2 plus 3 equals 5.** Open and close thumb and index finger, wiggle 3 fingers, extend 5 fingers. Repeat twice.

WRITTEN NUMBERS

MATERIALS: Lesson 6 Activity Sheet, pencils, crayons

- Say, **Now we are going to practice writing numbers. Here is a paper with the numbers 5 and 6. I would like you to trace over the numbers with your pencil. Do your best to stay on the dotted lines, but do not erase if you make a mistake. Just try it again! Under the dotted numbers is a place to copy the numbers on your own. Then write the number that tells us how many things are in the box.** Point to the pictures at the bottom of the page.

- As each child finishes, have him or her turn the paper over and say, **Write the missing number for these number sentences. If you do not know the missing number, then use your fingers to figure it out.**

Lesson 6 Activity Sheet

Lesson 7

Learning Goals	Materials
Establish behavior boundaries	**COPY**
Count to 30 orally	Hundreds Chart
Build numbers 1–7	Cardinality Chart
Number recognition 0–7	Subitizing Circle Cards (#1–28)
Count and sequence to 7	Number Sentence Cards: 5 − 2 = 3, 5 − 3 = 2, 5 − 1 = 4, 5 − 4 = 1
Before and after (1–7)	Lesson 7 Activity Sheet
Count to 7 on fingers	**GATHER**
Make numbers 1–7 on fingers	White board magnetic easel
Recognize quantities 0–7	3 interlocking blocks (of a different color than the prepared sticks of blocks) in a plastic bag
Story problems and number sentences: 5 family (minus)	Farm scene with 5 pig magnets
Perform number operations on fingers	Pencils without erasers and crayons
Write the numerals 6 and 7	**PREPARE**
Connect quantity to numerals 0–7	Put student names on Lesson 7 Activity Sheets.
Solve written number sentences: 5 family (minus)	25 interlocking blocks configured as 1 stick, 2 stick, 3 stick, 4 stick, 3 sets of 5 sticks
	Number Recognition Cards (0–7); *See Chapter 1 for instructions on making Number Recognition Cards.*
	Teacher Five Frame Mat with 5 two-colored magnetic dots; *See Chapter 1 for instructions on making the Five Frame Mat and two-colored dots.*
	Student Five Frame Trays with 5 two-colored magnetic dots

ESTABLISHING BEHAVIOR BOUNDARIES

- Say, **Show me you are ready to learn.** Wait for proper posture. **Exactly! That shows me you are ready!**

Number Sense Interventions Lesson 7   53

COUNTING WARM-UP

MATERIALS: Hundreds Chart, white board magnetic easel

- Put up the Hundreds Chart to use for reference when needed.
- Say, **Let's count to 30 this time, taking turns. I will start. 1** (point to the child to your left). If a child says the incorrect number, then say, **That was a good try but the next number is ___. Let's try again.** For each error, back up 2 children and repeat so the child has an opportunity to be successful. Take a turn yourself.

1	2	3	4	5	6	7	8	9	10
11	12	13	14	15	16	17	18	19	20
21	22	23	24	25	26	27	28	29	30
31	32	33	34	35	36	37	38	39	40
41	42	43	44	45	46	47	48	49	50
51	52	53	54	55	56	57	58	59	60
61	62	63	64	65	66	67	68	69	70
71	72	73	74	75	76	77	78	79	80
81	82	83	84	85	86	87	88	89	90
91	92	93	94	95	96	97	98	99	100

Hundreds Chart

MAGIC NUMBER ACTIVITIES (Magic Number is 7)

★ Cardinality

MATERIALS: Cardinality Chart, 25 interlocking blocks configured as 1 stick, 2 stick, 3 stick, 4 stick, 3 sets of 5 sticks; 3 interlocking blocks of another color

- Lay the Cardinality Chart on the table and say, **Here is our number chart. It has the numbers 1–10 across the bottom.** Run your finger across the numbers. Put the 1, 2, 3, 4, and 5 sticks on the table in a random arrangement.
- Say, **Let's put these blocks where they belong on the chart. (Child) , choose a stick of blocks and put them where they belong on the chart.** Continue until sticks 1–5 are on the chart.

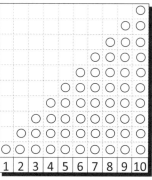

Cardinality Chart [number chart]

- Say, **Let's build numbers more than 5. I have 5 blocks in each of these sticks, like the 5 fingers on my hand.** Hold up one hand.
- Put 1 more block on the second stick while saying, **5 and 1 more is how many?** Wait for a response. Scaffold using the chart if necessary. Put down the stick of 6.
- Put 2 more blocks on the third stick while saying, **5 and 2 more is how many?** Wait for a response. Scaffold using the chart if necessary. Put down the stick of 7.
- Say, **That is right! 5 and 1 more is 6. 5 and 2 more is 7.** Gesture to the blocks while saying each number.
- Say, **As we go up the number chart** (gesture up the Cardinality Chart) **we add 1 block at a time. It is like climbing up steps, 1 at a time.** Use gestures from Lesson 2 while saying, **See, 5 and 1 more is 6, 6 and 1 more is 7, but 7 is also 5 and 2 more!** Point out the 5 and 2 more in the stick of 7.

★ Sequencing and Number Recognition 0–7

MATERIALS: Number Recognition Cards 0–7

- Say, **Say the numbers with me as I put them down. Do not go ahead of me. Ready? 1, 2, 3 ... 7!** Put the cards down horizontally as if building a number list. Orient the cards toward the children.
- Hold up the Number Recognition Card 7 and say, **What number is this?** Wait for a response. Say, **That is right. This is the number 7. Our Magic Number today is 7.**

Number Recognition Cards

- Say, **Let's play our game!** Follow error correction guidelines.
- Go around the group, showing each child a different number from 0 to 7 (not in order). Make sure all children can see the number. Go around the group three times.

★ Before and After

MATERIALS: Number Recognition Cards 1–7

- Put the Number Recognition Cards down in a row in random order and say, **Here are the numbers 1–7, but they are not in order.**
- Pull out the number 5. For the remainder of this exercise, call on individual children to choose the appropriate card.
- Begin by saying, **What number comes right before 5?** Point to the space before 5.
- Continue to alternately ask "before" and "after" questions until all the numbers are in sequence. Point to the spot where the missing number should be.
- When all numbers are in place, say, **Let's say our numbers to check. 1, 2, 3 ... 7. Does that count right?** Wait for a response.

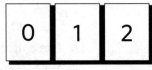
Number Recognition Cards

★ Finger Counting

- Say, **How many fingers are on one hand?** Wait for a response. **That is right! So if we count to 7, we can start with 5 and count up like this: 5, 6, 7.** Gesture one hand, then 2 fingers on the other hand, 1 finger at a time.
- Say, **Let's try counting to 7 again, starting with 5. Ready? 5, 6, 7.** Repeat.
- Say, **This is what 7 looks like on our fingers.** Demonstrate 7 all at once. **So if I say, "Show 7 on your fingers," what will you do?** Make sure every child shows 7 fingers as one hand and 2 fingers.

SUBITIZING QUANTITIES (1–7) ACTIVITIES

MATERIALS: Number Recognition Cards 0–7

★ Finger Automaticity

- Say, **Let's practice making 6 and 7 on our fingers quickly.**
- Say, **6 is 5 and 1 more.** Show one hand and 1 finger simultaneously. **Now you try it! Show me 6!**
- Say, **7 is 5 and 2 more.** Show one hand and 2 fingers simultaneously. **Now you try it! Show me 7!**
- Say, **Let's play our game.**
- Shuffle the Number Recognition Cards and show them one at a time. Go through the cards two times.
- Be sure that each child is holding up the correct number of fingers. If he or she is incorrect, then have him or her look at all the other children and correct him- or herself or let another child help him or her. Discourage counting and encourage automaticity.

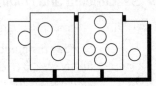
Number Recognition Cards

★ Recognizing Sets

MATERIALS: Subitizing Circle Cards (#1–28)

- Say, **Now I am going to hold up a card with some circles on it, and I want you to tell me how many circles are on the card. Try to tell me as fast as you can without counting. I will point to you for your turn, okay? Everyone else, say it in your mind.**

Subitizing Circle Cards

Number Sense Interventions Lesson 7   55

- Go through the Subitizing Circle Cards in random order. If a child gets a 4 or 5 card incorrect, then point out the smaller sets in the total. Go around the group four times.

COMPARING QUANTITIES

MATERIALS: Subitizing Circle Cards (#1–28)

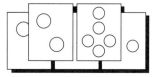
Subitizing Circle Cards

- Go through the Subitizing Circle Cards saying, **Listen carefully. Which has more?** Respond with, **That is right! ___ is more than ___.** Some children may naturally start saying the numbers but do not require it.
- Show 6 pairs.
- Say, **Now we are going to do something different. Listen carefully.**
- Go through the Subitizing Circle Cards saying, **Listen carefully. Which has less?** Respond with, **That is right! ___ is less than ___.** Some children may naturally start saying the numbers but do not require it.
- Show 6 pairs.

STORY PROBLEMS

MATERIALS: Number Sentence Cards, student Five Frame trays and two-colored dots, teacher Five Frame Mat and two-colored dots, white board magnetic easel, farm scene with 5 pig magnets

- Say, **Let's go to the farm!** Put the farm scene on the white board magnetic easel. Model each story on the farm scene as you tell it.

5 – 2 = 3

Number Sentence Card

- Say, **The farmer bought 5 pigs at the market. When he was putting them into the pen, 2 ran away. How many pigs are still in the pen?** Wait for a response. Say, **That is right. 5 pigs take away 2 pigs leaves 3 pigs.** Gesture to the picture.
- Hand out the Student Five Frame Trays with dots.

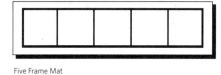
Five Frame Mat

- Say, **Let's show that story on our frames. First make all 5 dots red. Now turn over 2 for the pigs that ran away. How many are still red?** Wait for a response. **That is right. 5 minus 2 equals 3.** Help children who are struggling.
- Put the Number Sentence Cards on the table.
- Say, **Which number sentence goes with this story?** Call on a volunteer to choose the correct card.

5 – 3 = 2

- Say, **The farmer bought 5 pigs at the market. He sold 3 of the pigs to his neighbor. How many pigs did the farmer have left?** Wait for a response. Say, **That is right. 5 pigs take away 3 pigs leaves 2 pigs.** Gesture to the picture.
- Say, **Let's show that story on our frames. First make all 5 dots red. Now turn over 3 for the pigs that were sold. How many are still red?** Wait for a response. **That is right. 5 minus 3 equals 2.** Help children who are struggling.
- Say, **Which number sentence goes with this story?** Call on a volunteer to choose the correct card.

$5 - 1 = 4$

- Say, **The farmer put 5 pigs in the pen. 1 got sick so he put it in the barn to stay warm. How many pigs were left in the pen?** Wait for a response. Say, **That is right. 5 pigs take away 1 pig leaves 4 pigs.**
- Say, **Let's show that story on our frames. First make all 5 dots red. Now turn over 1 for the pig that got sick. How many are still red?** Wait for a response. **That is right. 5 minus 1 equals 4.** Help children who are struggling.
- Say, **Which number sentence goes with this story?** Call on a volunteer to choose the correct card.

$5 - 4 = 1$

- Hold up the 5 − 4 = 1 card. Say, **Who would like to tell a story for this card?**
- Choose a child to tell the story and guide him or her if he or she needs help. Model on the Five Frame Mat.

NUMBER SENTENCES ON FINGERS

MATERIALS: Number Sentence Cards

- Say, **We can also show these number sentences on our fingers.**

$5 - 1 = 4$

- Hold up the 5 − 1 = 4 card. Say, **This says 5 minus 1 equals 4. 5 minus 1 equals 4.** While saying number sentence the second time, show 5 fingers on the table, push your pinky off the table, and wiggle 4 remaining fingers as you say, "4." **Now you try.** Repeat twice.

$5 - 2 = 3$

- Hold up the 5 − 2 = 3 card. Say, **This says 5 minus 2 equals 3. 5 minus 2 equals 3.** While saying number sentence the second time, show 5 fingers on the table, push your pinky and ring finger off the table, and wiggle 3 remaining fingers as you say, "3." **Now you try.** Repeat twice.

$5 - 3 = 2$

- Hold up the 5 − 3 = 2 card. Say, **This says 5 minus 3 equals 2. 5 minus 3 equals 2.** While saying number sentence the second time, show 5 fingers on the table; push your pinky, ring finger, and middle finger off the table; and wiggle 2 remaining fingers as you say, "2." **Now you try.** Repeat twice.

$5 - 4 = 1$

- Hold up the 5 − 4 = 1 card. Say, **This says 5 minus 4 equals 1. 5 minus 4 equals 1.** While saying number sentence the second time, push all 4 fingers off the table and wiggle your thumb as you say, "1." **Now you try.** Repeat twice.

Number Sense Interventions Lesson 8

WRITTEN NUMBERS

MATERIALS: Lesson 7 Activity Sheet, pencils, crayons

- Say, **Now we are going to practice writing numbers. Here is a paper with the numbers 6 and 7. I would like you to trace over the numbers with your pencil. Do your best to stay on the dotted lines, but do not erase if you make a mistake. Just try it again! Under the dotted numbers is a place to copy the numbers on your own. Then write the number that tells us how many things are in the box.** Point to the pictures at the bottom of the page.

- As each child finishes, have him or her turn the paper over and say, **Write the missing number for these number sentences. If you do not know the missing number, then use your fingers to figure it out.**

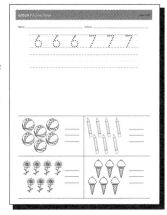

Lesson 7 Activity Sheet

Lesson 8

Learning Goals	Materials
Establish behavior boundaries	**COPY**
Count to 30 orally	Hundreds Chart
Build numbers 1-8	Cardinality Chart
Number recognition 0-8	Subitizing Circle Cards (#1-28)
Count and sequence to 8	Number Sentence Cards: 5 + 1 = 6, 5 + 2 = 7, 5 + 3 = 8
Before and after (1-8)	Lesson 8 Activity Sheet
Count to 8 on fingers	**GATHER**
Make numbers 1-8 on fingers	White board magnetic easel
Recognize quantities 0-8	6 interlocking blocks (of a different color than the prepared sticks of blocks) in a plastic bag
Make 6, 7, 8 on Ten Frame Mat	Farm scene with 8 duck magnets
Story problems and number sentences: 5 + 1, 5 + 2, 5 + 3	Pencils without erasers and crayons
Perform number operations on fingers	**PREPARE**
Write the numerals 7 and 8	Put student names on Lesson 8 Activity Sheets.
Connect quantity to numerals 0-8	30 interlocking blocks in sticks configured as 1 stick, 2 stick, 3 stick, 4 stick, 4 sets of 5 sticks
Solve written number sentences: 5 family (plus and minus)	Number Recognition Cards (0-8); See Chapter 1 for instructions on making Number Recognition Cards.
	Teacher Ten Frame Mat with 8 two-colored dots; See Chapter 1 for instructions on making the Ten Frame Mat and two-colored dots.
	Student Ten Frame Trays with 8 two-colored dots
	Cover for the Ten Frame

ESTABLISHING BEHAVIOR BOUNDARIES

- Say, **Show me you are ready to learn.** Wait for proper posture. **Exactly! That shows me you are ready!**

COUNTING WARM-UP

MATERIALS: Hundreds Chart, white board magnetic easel

- Put up the Hundreds Chart to use for reference when needed.
- Say, **Let's count to 30 again, taking turns. I will start. 1** (point to the child to your left). If a child says the incorrect number, then say, **That was a good try but the next number is ___. Let's try again.** For each error, back up 2 children and repeat so the child has an opportunity to be successful. Take a turn yourself.

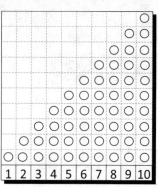

Hundreds Chart

MAGIC NUMBER ACTIVITIES (Magic Number is 8)

★ Cardinality

MATERIALS: Cardinality Chart, 30 interlocking blocks configured as 1 stick, 2 stick, 3 stick, 4 stick, 4 sets of 5 sticks, 6 interlocking blocks of another color

- Lay the Cardinality Chart on the table and say, **Here is our number chart. It has the numbers 1–10 across the bottom.** Run your finger across the numbers. Put the 1, 2, 3, 4, and 5 sticks on the table in a random arrangement.
- Say, **Let's put these blocks where they belong on the chart. _(Child)_, choose a stick of blocks and put them where they belong on the chart.** Continue until sticks 1–5 are on the chart.
- Say, **Let's build numbers more than 5. I have 5 blocks in each of these sticks, like the 5 fingers on my hand.** Hold up one hand.
- Put 1 more block on the second stick while saying, **5 and 1 more is how many?** Wait for the response. Scaffold using the chart if necessary. Put down the stick of 6.
- Put 2 more blocks on the third stick while saying, **5 and 2 more is how many?** Wait for a response. Scaffold using the chart if necessary. Put down the stick of 7.
- Put 3 more blocks on the third stick while saying, **5 and 3 more is how many?** Wait for a response. Scaffold using the chart if necessary. Put down the stick of 8.
- Say, **That is right! 5 and 1 more is 6. 5 and 2 more is 7. 5 and 3 more is 8.** Gesture to the blocks while saying each number.
- Say, **As we go up the number list** (gesture up the Cardinality Chart), **we add 1 block at a time. It is like climbing up steps, 1 at a time.** Use gestures from Lesson 2 while saying, **See, 1 and 1 more is 2, 5 and 1 more is 6, 6 and 1 more is 7, 7 and 1 more is 8!**

Cardinality Chart [number chart]

★ Sequencing and Number Recognition 0–8

MATERIALS: Number Recognition Cards 0–8

- Say, **Say the numbers with me as I put them down. Do not go ahead of me. Ready? 1, 2, 3 … 8!** Put the cards down horizontally as if building a number list. Orient the cards toward the children.
- Hold up the Number Recognition Card 8 and say, **What number is this?** Wait for a response.
- Say, **That is right. This is the number 8. Our Magic Number today is 8.**

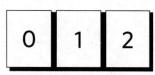

Number Recognition Cards

Number Sense Interventions Lesson 8 59

- Say, **Let's play our game!** Follow the error correction guidelines.
- Go around the group, showing each child a different number from 0 to 8 (not in order). Make sure all children can see the number. Go around the group three times.

★ Before and After

MATERIALS: Number Recognition Cards 1–8

- Put the Number Recognition Cards down in a row in random order and say, **Here are the numbers 1–8, but they are not in order.**
- Pull out the number 4. For the remainder of this exercise, call on individual children to choose the appropriate card.
- Begin by saying, **What number comes right before 4?** Point to the space before 4.
- Continue to alternately ask "before" and "after" questions until all the numbers are in sequence. Point to the spot where the missing number should be.
- When all numbers are in place, say, **Let's say our numbers to check. 1, 2, 3 … 8. Does that count right?** Wait for a response.

Number Recognition Cards

★ Finger Counting

- Say, **How many fingers are on one hand?** Wait for a response. **That is right! So if we count to 8, we can start with 5 and count up like this: 5, 6, 7, 8.** Gesture one hand, then 3 fingers on the other hand, one finger at a time.
- Say, **Let's try counting to 8 again, starting with 5. Ready? 5, 6, 7, 8.** Repeat.
- Say, **This is what 8 looks like on our fingers.** Demonstrate 8 all at once. **So if I say show 8 on your fingers, what will you do?** Make sure every child shows 8 fingers as one hand and 3 fingers.

SUBITIZING QUANTITIES (1–8) ACTIVITIES

★ Finger Automaticity

MATERIALS: Number Recognition Cards

- Say, **Let's practice making 6, 7, and 8 on our fingers quickly.**
- Say, **6 is 5 and 1 more.** Show one hand and 1 finger simultaneously. **Now you try it! Show me 6!**
- Say, **7 is 5 and 2 more.** Show one hand and 2 fingers simultaneously. **Now you try it! Show me 7!**
- Say, **8 is 5 and 3 more.** Show one hand and 3 fingers simultaneously. **Now you try it! Show me 8!**
- Say, **Let's play our game.**
- Shuffle the Number Recognition Cards and show them one at a time. Go through the cards two times. Correct errors.

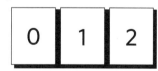
Number Recognition Cards

★ Recognizing Sets

MATERIALS: Subitizing Circle Cards #1–28

- Say, **Now I am going to hold up a card with some circles on it, and I want you to tell me how many circles are on the card. Try to tell me as fast as you can without counting. I will point to you for your turn, okay? Everyone else, say it in your mind.**
- Go through the Subitizing Circle Cards in random order. If a child gets a 4 or 5 card incorrect, then point out the smaller sets in the total. Go around the group four times.

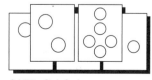
Subitizing Circle Cards

COMPARING QUANTITIES

MATERIALS: Subitizing Circle Cards #1-28

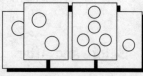
Subitizing Circle Cards

- Go through the Subitizing Circle Cards saying, **Listen carefully. Which has more?** Respond with, **That is right! ___ is more than ___.** Some children may naturally start saying the numbers but do not require it.
- Show 6 pairs.
- Say, **Now we are going to do something different. Listen carefully.**
- Go through the Subitizing Circle Cards saying, **Listen carefully. Which has less?** Respond with, **That is right! ___ is less than ___.** Some children may naturally start saying the numbers but do not require it.
- Show 6 pairs.

TEN FRAME ACTIVITIES

MATERIALS: Teacher Ten Frame Mat with 8 two-colored dots, cover for one Five Frame, white board magnetic easel

★ Making 5, 6, 7, 8

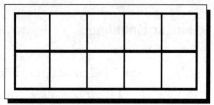
Ten Frame Mat

- Put 5 red dots in the top row and 3 yellow dots in the bottom row.
- Put the Ten Frame Mat on the white board magnetic easel with 5 dots covered and say, **Here is a Five Frame with 5 dots. But what if I want to show 6 or 7 dots? I need more boxes. I will add another Five Frame.** Uncover.
- Say, **5 and 5.** Point to the 2 sets of blocks. **Now we have enough rows for two hands, not just one.** Show two hands, one at a time. **5 and 5.**
- Say, **We know how to make 6 on our fingers. 5 and 1.** Demonstrate on fingers. **It looks the same on our frame. 5 and 1.** Put the dots on the Ten Frame Mat.
- Say, **See, 5 and 1; 5, 6.** Draw your finger across the row of 5 when saying, "5." Point to the single dot in the bottom row when saying, "1" or "6."
- Say, **We know how to make 7 on our fingers. 5 and 2.** Demonstrate on fingers. **It looks the same on our frame. 5 and 2.** Put another dot on the Ten Frame Mat.
- Say, **See, 5 and 2; 5, 6, 7.** Draw your finger across the row of 5 when saying, "5," and across the 2 dots in the bottom row when saying, "2" (touch each individual dot when saying, "6, 7").
- Say, **We know how to make 8 on our fingers. 5 and 3.** Demonstrate on fingers. **It looks the same on our frame. 5 and 3.** Put another dot on the Ten Frame Mat.
- Say, **See, 5 and 3; 5, 6, 7, 8.** Draw your finger across the row of 5 when saying, "5," and across the 3 dots in the bottom row when saying, "3" (touch each individual dot when saying, "6, 7, 8").

★ Ten Frame Game

MATERIALS: Student Ten Frame Trays with 8 two-colored dots, Number Recognition Cards 5-8

- Hand out the Ten Frame Trays and say, **Now it is your turn! I am going to show you a number from 5 to 8, and I want you to quickly make it on your Ten Frame and show me when you are finished. Keep the top dots red and the bottom dots yellow. Ready?**

- Hold up a Number Recognition Card in random order and have children build it on their Ten Frames. They should hold up their Ten Frame so that it faces you when they are finished.
- If a child is incorrect, then have him or her count, but make sure he or she counts starting at 5, not 1.
- Collect the Ten Frame Trays.

STORY PROBLEMS

MATERIALS: Number Sentence Cards, farm scene with 8 duck magnets, white board magnetic easel

- Say, **Let's go to the farm!** Put the farm scene on the white board magnetic easel. Put the Number Sentence Cards on the table.
- Model each story on the farm scene as you tell it.

5 + 1 = 6

- Say, **Listen to this story. There are 5 ducks in the pond.** Show the 5 ducks on your fingers. **Then 1 more flies into the pond.** Show the 1 duck on your other hand. **How many ducks are in the pond now?**
- Make sure everyone has his or her hand up and say, **What number is this? Do you know without counting?** Wait for a response.
- Say, **Which number sentence goes with this story: 5 ducks and 1 more duck?** Call on a child to choose the correct Number Sentence Card. If he or she chooses the incorrect card, then say, **This card says 5 and __more. Which card says 5 and 1 more?**

5 + 2 = 7

- Say, **Listen to this story. There are some ducks at the farm. 5 are in the pond. Show me 5 on your fingers. 2 are in the grass. Show me 2 on your other hand. How many ducks are there altogether?**
- Make sure everyone has his or her hand up and say, **What number is this? Do you know without counting?** Wait for a response.
- Say, **Which number sentence goes with this story: 5 ducks and 2 more ducks?** Call on a child to choose the correct Number Sentence Card. If he or she chooses the incorrect card, then say, **This card says 5 and __ more. Which card says 5 and 2 more?**

5 + 3 = 8

- Say, **Listen to this story. There are 5 ducks in the grass. Show me 5 ducks on your fingers. The farmer brings 3 more ducks home from the market to join them. Show me 3 more on your other hand. How many ducks are in the grass now?**
- Make sure everyone has his or her hand up and say, **What number is this? Do you know without counting?** Wait for a response.
- Say, **Which number sentence goes with this story: 5 ducks and 3 more ducks?** Call on a child to choose the correct Number Sentence Card. If he or she chooses the incorrect card, then say, **This card says 5 and __ more. Which card says 5 and 3 more?**

NUMBER SENTENCES ON FINGERS

MATERIALS: Number Sentence Cards

- Pick up the Number Sentence Cards and say, **Now we are going to show these number sentences on our fingers.**
- Say, **I will hold up a card and you make it on your fingers. Remember, we can show 5 on one hand by showing all our fingers.** Show one hand. **We do not need to count!**
- Hold up the 5 + 1 = 6 card. Say, **Let's try this one. It says 5 plus 1. Let's show 5 plus 1.** Show one hand and 1 finger.
- Say, **I am going to give you each some number sentences to show on your fingers.** Show each child two Number Sentence Cards to perform on their fingers.

$5 + 1 = 6$

Number Sentence Card

WRITTEN NUMBERS

MATERIALS: Lesson 8 Activity Sheet, pencils, crayons

- Say, **Now we are going to practice writing numbers. Here is a paper with the numbers 7 and 8. I would like you to trace over the numbers with your pencil. Do your best to stay on the dotted lines, but do not erase if you make a mistake. Just try it again! Under the dotted numbers is a place to copy the numbers on your own. Then write the number that tells us how many things are in the box.** Point to the pictures at the bottom of the page.
- As each child finishes, have him or her turn the paper over and say, **Write the missing number for these number sentences. If you do not know the missing number, then use your fingers to figure it out.**

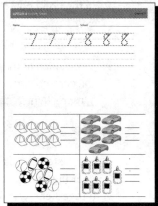

Lesson 8 Activity Sheet

Lesson 9

Learning Goals

Establish behavior boundaries
Count to 30 orally
Build numbers 1-9
Number recognition 0-9
Count and sequence to 9
Before and after (1-9)
Count to 9 on fingers
Make numbers 1-9 on fingers
Recognize quantities 0-9
Represent 5-9 on Ten Frame
Story problems and number sentences: 5 + 1, 5 + 2, 5 + 3, 5 + 4
Perform number operations on fingers
Write the numerals 8 and 9
Solve written number sentences: 5 + 1, 5 + 2, 5 + 3, 5 + 4

Materials

COPY

Hundreds Chart
Cardinality Chart
Subitizing Circle Cards #1-28
Number Sentence Cards: 5 + 1 = 6, 5 + 2 = 7, 5 + 3 = 8, 5 + 4 = 9
Lesson 9 Activity Sheet

GATHER

White board magnetic easel
10 interlocking blocks (of a different color than the prepared sticks of blocks) in a plastic bag
Pencils without erasers and crayons

PREPARE

Put student names on Lesson 9 Activity Sheets.
35 interlocking blocks configured as 1 stick, 2 stick, 3 stick, 4 stick, 5 sets of 5 sticks
Number Recognition Cards (0-9); *See Chapter 1 for instructions on making Number Recognition Cards.*
Teacher Ten Frame Mat with 9 two-colored dots; *See Chapter 1 for instructions on making the Ten Frame Mat and two-colored dots.*
Student Ten Frame Trays with 9 two-colored magnetic dots

ESTABLISHING BEHAVIOR BOUNDARIES

- Say, **Show me you are ready to learn.** Wait for proper posture. **Exactly! That shows me you are ready!**

COUNTING WARM-UP

MATERIALS: Hundreds Chart, white board magnetic easel

- Put up the Hundreds Chart to use for reference when needed.
- Say, **Let's count to 30 again, taking turns. I will start. 1** (point to the child to your left). If a child says the incorrect number, then say, **That was a good try but the next number is ____. Let's try again.** For each error, back up 2 children and repeat so the child has an opportunity to be successful. Take a turn yourself.

1	2	3	4	5	6	7	8	9	10
11	12	13	14	15	16	17	18	19	20
21	22	23	24	25	26	27	28	29	30
31	32	33	34	35	36	37	38	39	40
41	42	43	44	45	46	47	48	49	50
51	52	53	54	55	56	57	58	59	60
61	62	63	64	65	66	67	68	69	70
71	72	73	74	75	76	77	78	79	80
81	82	83	84	85	86	87	88	89	90
91	92	93	94	95	96	97	98	99	100

Hundreds Chart

MAGIC NUMBER ACTIVITIES (The Magic Number is 9)

★ Cardinality

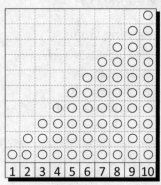

Cardinality Chart [number chart]

MATERIALS: Cardinality Chart, 35 interlocking blocks configured as 1 stick, 2 stick, 3 stick, 4 stick, 5 sets of 5 sticks, 10 interlocking blocks of another color

- Lay the Cardinality Chart on the table and say, **Here is our number chart. It has the numbers 1–10 across the bottom.** Run your finger across the numbers. Put the 1, 2, 3, 4, and 5 sticks on the table in a random arrangement.
- Say, **Let's put these blocks where they belong on the chart. (Child), choose a stick of blocks and put them where they belong on the chart.** Continue until sticks 1–5 are on the chart.
- Say, **Let's build numbers more than 5. I have 5 blocks in each of these sticks, like the 5 fingers on my hand.** Hold up one hand.
- Put 1 more block on the second stick while saying, **5 and 1 more is how many?** Wait for a response. Scaffold using the chart if necessary. Put down the stick of 6.
- Repeat with 5 + 2, 5 + 3, 5 + 4.
- Say, **That is right! 5 and 1 more is 6. 5 and 2 more is 7. 5 and 3 more is 8. 5 and 4 more is 9.** Gesture to the blocks while saying each number.
- Say, **As we go up the number chart** (gesture up the Cardinality Chart) **we add 1 block at a time. It is like climbing up steps, 1 at a time.** Use gestures from Lesson 2 while saying, **See, 5 and 1 more is 6, 6 and 1 more is 7, 7 and 1 more is 8, 8 and 1 more is 9!**

★ Sequencing and Number Recognition 0–9

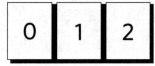

Number Recognition Cards

MATERIALS: Number Recognition Cards 0–9

- Say, **Say the numbers with me as I put them down. Do not go ahead of me. Ready? 1, 2, 3 … 9!** Put the cards down horizontally as if building a number list. Orient the cards toward the children.
- Hold up the Number Recognition Card 9 and say, **What number is this?** Wait for a response.
- Say, **That is right. This is the number 9. Our Magic Number today is 9.**
- Say, **Let's play our game!**
- Go around the group, showing each child a different number from 0 to 9 (not in order). Make sure all children can see the number. Go around the group three times.

★ Before and After

Number Recognition Cards

MATERIALS: Number Recognition Cards 1–9

- Put the Number Recognition Cards down in a row in random order and say, **Here are the numbers 1–9, but they are not in order.**
- Pull out the number 6. For the remainder of this exercise, call on individual children to choose the appropriate card.
- Begin by saying, **What number comes right before 6?** Point to the space before 6.
- Continue to alternately ask "before" and "after" questions until all the numbers are in sequence. Point to the spot where the missing number should be.
- When all numbers are in place, say, **Let's say our numbers to check: 1, 2, 3 … 9. Does that count right?** Wait for a response.

Number Sense Interventions Lesson 9   65

★ Finger Counting

- Say, **How many fingers are on one hand?** Wait for a response. **That is right! So if we count to 9, we can start with 5 and count up like this: 5, 6, 7, 8, 9.** Gesture one hand, then 4 fingers on the other hand, one finger at a time.
- Say, **Let's try counting to 9 again, starting with 5. Ready? 5, 6, 7, 8, 9.** Repeat.
- Say, **This is what 9 looks like on our fingers.** Demonstrate 9 fingers all at once. **So if I say show 9 on your fingers, what will you do?** Make sure every child shows 9 fingers as one hand and 4 fingers.

SUBITIZING QUANTITIES (1-9) ACTIVITIES

★ Finger Automaticity

MATERIALS: Number Recognition Cards 1-9

- Say, **Let's practice making 6, 7, 8, and 9 on our fingers quickly.**
- Say, **6 is 5 and 1 more.** Show one hand and 1 finger simultaneously. **Now you try it! Show me 6!**
- Say, **7 is 5 and 2 more.** Show one hand and 2 fingers simultaneously. **Now you try it! Show me 7!**
- Say, **8 is 5 and 3 more.** Show one hand and 3 fingers simultaneously. **Now you try it! Show me 8!**
- Say, **9 is 5 and 4 more.** Show one hand and 4 fingers simultaneously. **Now you try it! Show me 9!**
- Say, **Let's play our game.**
- Shuffle the Number Recognition Cards and show them one at a time. Go through the cards two times. Correct errors.

Number Recognition Cards

★ Recognizing Sets

MATERIALS: Subitizing Circle Cards (#1-28)

- Say, **Now I am going to hold up a card with some circles on it, and I want you to tell me how many circles are on the card. Try to tell me as fast as you can without counting. I will point to you for your turn, okay? Everyone else, say it in your mind.**
- Go through the Subitizing Circle Cards in random order. If a child gets a 4 or 5 card incorrect, then point out the smaller sets in the total. Go around the group four times.

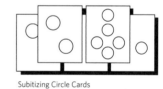
Subitizing Circle Cards

COMPARING QUANTITIES

MATERIALS: Subitizing Circle Cards (#1-28)

- Go through the Subitizing Circle Cards saying, **Listen carefully. Which has more?** Respond with, **That is right! ___ is more than ___.** Some children may naturally start saying the numbers but do not require it.
- Show 6 pairs.
- Say, **Now we are going to do something different. Listen carefully.**
- Go through the Subitizing Circle Cards saying, **Listen carefully. Which has less?** Respond with, **That is right! ___ is less than ___.** Some children may naturally start saying the numbers but do not require it.
- Show 6 pairs.

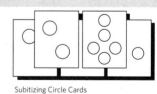
Subitizing Circle Cards

66 Jordan and Dyson

TEN FRAME ACTIVITIES

MATERIALS: Teacher Ten Frame Mat with 9 two-colored dots, white board magnetic easel

★ Making 5, 6, 7, 8, 9

- Put 5 red dots in the top row and 4 yellow dots in the bottom row.
- Put the Ten Frame Mat on the white board magnetic easel and say, **Here is a Ten Frame. 5 and 5.** Point to the 2 sets of blocks. **5 blocks for this hand and 5 blocks for this hand.** Show each hand, one at a time. **5 and 5.**
- Say, **We know how to make 6 on our fingers. 5 and 1.** Demonstrate on fingers. **It looks the same on our frame. 5 and 1.** Put the dots on the Ten Frame Mat.
- Say, **See, 5 and 1; 5, 6.** Draw your finger across the row of 5 when saying, "5," point to the single dot in the bottom row when saying, "1 or 6."
- Repeat for 5 + 2, 5 + 3, 5 + 4.

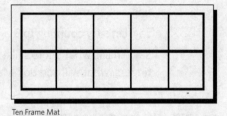
Ten Frame Mat

★ Ten Frame Game

MATERIALS: Student Ten Frame Trays with 9 two-colored magnetic dots, Number Recognition Cards 5–9

- Hand out the Student Ten Frame Trays and say, **Now it is your turn! I am going to show you a number from 5 to 9, and I want you to quickly make it on your Ten Frame and show me when you are finished. Keep the top top dots red and the bottom dots yellow. Ready?**
- Hold up a Number Recognition Card in random order and have children build it on their Ten Frames. They should hold up their Ten Frames so that it faces you when they are finished.
- If a child is incorrect, then have him or her count, but make sure he or she counts starting at 5, not 1.
- Collect the Ten Frame Trays.

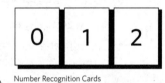
Number Recognition Cards

STORY PROBLEMS

MATERIALS: Number Sentence Cards

- Say, **I am going to tell you some stories. Make the stories on your fingers.** Put the Number Sentence Cards on the table.

5 + 1 = 6

- Say, **Listen to this story. Mike has 5 markers. Show me Mike's markers on your fingers. His sister gives him 1 more marker. Show 1 more marker on your fingers. How many markers does Mike have now? Tell me without counting!** Wait for a response.
- Say, **Which number sentence goes with this story: 5 markers and 1 more marker?** Call on a child to choose the correct card. If he or she chooses the incorrect card, then say, **This card says 5 and __ more. Which card says 5 and 1 more?**

5 + 1 = 6
Number Sentence Card

5 + 2 = 7

- Say, **Listen to this story. Christine has 5 pencils. Show me Christine's pencils using your fingers. Jill gives her 2 more pencils. Show 2 more pencils on your fingers. How many pencils does Christine have now? Tell me without counting!** Wait for a response.

5 + 2 = 7
Number Sentence Card

Number Sense Interventions Lesson 9   67

- Say, **Which number sentence goes with this story: 5 pencils and 2 more pencils?** Call on a child to choose the correct card. If he or she chooses the incorrect card, then say, **This card says 5 and __ more. Which card says 5 and 2 more?**

5 + 3 = 8

- Say, **Listen to this story. Juan picked 5 apples. Show me Juan's apples on your fingers. He picked 3 more apples. Show 3 more apples on your fingers. How many apples does Juan have now? Tell me without counting!** Wait for a response.

5 + 3 = 8
Number Sentence Card

- Say, **Which number sentence goes with this story: 5 apples and 3 more apples?** Call on a child to choose the correct card. If he or she chooses the incorrect card, then say, **This card says 5 and __ more. Which card says 5 and 3 more?**

5 + 4 = 9

- Say, **Listen to this story. Demiah had 5 DVDs downstairs. Show me Demiah's DVDs on your fingers. She had 4 DVDs upstairs. Show 4 more DVDs on your fingers. How many DVDs does Demiah have altogether? Tell me without counting!** Wait for a response.

5 + 4 = 9
Number Sentence Card

- Say, **Which number sentence goes with this story: 5 DVDs and 4 more DVDs?** Call on a child to choose the correct card. If he or she chooses the incorrect card, then say, **This card says 5 and __ more. Which card says 5 and 4 more?**

NUMBER SENTENCES ON FINGERS

MATERIALS: Number Sentence Cards

- Pick up the Number Sentence Cards and say, **Now we are going to show these number sentences on our fingers.**

5 + 1 = 6
Number Sentence Card

- Say, **I will hold up a card and you make it on your fingers. Remember, we can show 5 on one hand by showing all our fingers.** Show one hand. **We do not need to count!**
- Say, **Let's try this one. It says 5 plus 1. Let's show 5 plus 1.** Show one hand and 1 finger.
- Say, **I am going to give you each some number sentences to show on your fingers.** Show each child two Number Sentence Cards to perform on their fingers.

WRITTEN NUMBERS

MATERIALS: Lesson 9 Activity Sheet, pencils without erasers, crayons

- Say, **Now we are going to practice writing numbers. Here is a paper with the numbers 8 and 9. I would like you to trace over the numbers with your pencil. Do your best to stay on the dotted lines, but do not erase if you make a mistake. Just try it again! Under the dotted numbers is a place to copy the numbers on your own. Then write the number that tells us how many things are in the box.** Point to the pictures at the bottom of the page.
- As each child finishes, have him or her turn the paper over and say, **Write the missing number for these number sentences. If you do not know the missing number, then use your fingers to figure it out.**

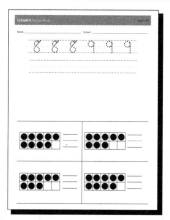

Lesson 9 Activity Sheet

Lesson 10

Learning Goals

Establish behavior boundaries
Count to 40 orally
Build numbers 1–10
Number recognition 0–10
Count and sequence to 10
Before and after (1–10)
Count to 10 on fingers
Make numbers 1–10 on fingers
Recognize quantities 0–10
Represent 5–10 on Ten Frame
Story problems and number sentences: 5 – 1, 5 – 2, 5 – 3, 5 – 4, 5 – 5
Perform number operations on fingers
Write the numerals 9 and 10
Connect quantity to numerals 0–10
Solve written number sentences: 5 – 1, 5 – 2, 5 – 3, 5 – 4, 5 – 5

Materials

COPY

Hundreds Chart
Cardinality Chart
Ten Frame Flash Cards
Number Sentence Cards: 5 – 1 = 4, 5 – 2 = 3, 5 – 3 = 2, 5 – 4 = 1, 5 – 5 = 0, 5 + 1 = 6, 5 + 2 = 7, 5 + 3 = 8, 5 + 4 = 9, 5 + 5 = 10
Lesson 10 Activity Sheet

GATHER

White board magnetic easel
15 interlocking blocks (of a different color than the prepared sticks of blocks) in a plastic bag
Pencils without erasers and crayons

PREPARE

Put student names on Lesson 10 Activity Sheets.
30 interlocking blocks in sticks configured as 1 stick, 2 stick, 3 stick, 4 stick, 6 sets of 5 sticks
Number Recognition Cards (0–10); *See Chapter 1 for instructions on making Number Recognition Cards.*
Teacher Ten Frame Mat with 10 two-colored dots; *See Chapter 1 for instructions on making the Ten Frame Mat and two-colored dots.*
Student Ten Frame Trays with 10 two-colored dots

ESTABLISHING BEHAVIOR BOUNDARIES

- Say, **Show me you are ready to learn.** Wait for proper posture. **Exactly! That shows me you are ready!**

COUNTING WARM-UP

MATERIALS: Hundreds Chart, white board magnetic easel

- Put up the Hundreds Chart to use for reference when needed.
- Say, **Let's count to 40 this time, but we will start counting at 11. I will start. 11** (point to the child to your right). If a child says the incorrect number, then say, **That was a good try, but the next number is 12. Let's try again.** For each error, back up 2 children and repeat so the child has an opportunity to be successful. Take a turn yourself.

Hundreds Chart

Number Sense Interventions Lesson 10   69

MAGIC NUMBER ACTIVITIES (Magic Number is 10)

★ Cardinality

MATERIALS: Cardinality Chart, 30 interlocking blocks configured as 1 stick, 2 stick, 3 stick, 4 stick, 6 sets of 5 sticks; 15 interlocking blocks of another color

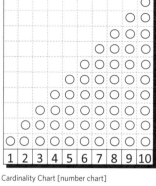
Cardinality Chart [number chart]

- Lay the Cardinality Chart on the table and say, **Here is our number chart. It has the numbers 1-10 across the bottom.** Run your finger across the numbers. Put the 1, 2, 3, 4, and 5 sticks on the table in a random arrangement.
- Say, **Let's put these blocks where they belong on the chart. (Child), choose a stick of blocks and put them where they belong on the chart.** Continue until sticks 1-5 are on the chart.
- Say, **Let's build numbers more than 5. I have 5 blocks in each of these sticks, like the 5 fingers on my hand.** Hold up one hand.
- Put 1 more block on the second stick while saying, **5 and 1 more is how many?** Wait for a response. Scaffold using the chart if necessary. Put down the stick of 6.
- Repeat with 5 + 2, 5 + 3, 5 + 4.
- Say, **Now our chart is finished!**
- Say, **As we go up the number list** (gesture up the Cardinality Chart), **we add 1 block at a time. It is like climbing up steps, 1 at a time.** Use gestures from Lesson 2 while saying, **See, 1 and 1 more is 2** and so forth up to **9 and 1 more is 10.**

★ Sequencing and Number Recognition 0-10

MATERIALS: Number Recognition Cards 0-10

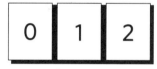
Number Recognition Cards

- Say, **Say the numbers with me as I put them down. Do not go ahead of me. Ready? 1, 2, 3 ... 10!** Put the cards down horizontally as if building a number list. Orient the cards toward the children.
- Hold up the Number Recognition Card 10 and say, **What number is this?** Wait for a response.
- Say, **That is right. This is the number 10. Our Magic Number today is 10.**
- Say, **Let's play our game!**
- Go around the group, showing each child a different number from 1 to 10 (not in order). Make sure all children can see the number. Go around the group three times.

★ Before and After

MATERIALS: Number Recognition Cards 1-10

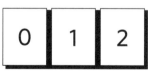
Number Recognition Cards

- Put the Number Recognition Cards down in a row in random order and say, **Here are the numbers 1-10, but they are not in order.**
- Pull out the number 5. For the remainder of this exercise, call on individual children to choose the appropriate card.
- Begin by saying, **What number comes right before 5?** Point to the space before 5.
- Continue to alternately ask "before" and "after" questions until all the numbers are in sequence. Point to the spot where the missing number should be.
- When all numbers are in place, say, **Let's say our numbers to check: 0, 1, 2, 3 ... 10. Does that count right?** Wait for a response.

★ Finger Counting

- Say, **How many fingers are on one hand?** Wait for a response. **That is right! So if we count to 10, we can start with 5 and count up like this: 5, 6, 7, 8, 9, 10!** Gesture one hand, then 5 fingers on the other hand, one at a time.
- Say, **Let's try counting to 10 again, starting with 5. Ready? 5, 6, 7, 8, 9, 10.** Repeat.
- Say, **This is what 10 looks like on our fingers. We used all our fingers!** Demonstrate 10 all at once. **So if I say show 10 on your fingers, what will you do?** Make sure every child shows 10 fingers as two hands.

SUBITIZING QUANTITIES (1–10) ACTIVITIES

★ Finger Automaticity

MATERIALS: Number Recognition Cards 1–10

- Say, **Let's practice making numbers on our fingers quickly.**
- Say, **6 is 5 and 1 more.** Show one hand and 1 finger simultaneously. **Now you try it! Show me 6!**
- Say, **7 is 5 and 2 more.** Show one hand and 2 fingers simultaneously. **Now you try it! Show me 7!**
- Say, **8 is 5 and 3 more.** Show one hand and 3 fingers simultaneously. **Now you try it! Show me 8!**
- Say, **9 is 5 and 4 more.** Show one hand and 4 fingers simultaneously. **Now you try it! Show me 9!**
- Say, **10 is 5 and 5 more.** Show one hand and 5 fingers simultaneously. **Now you try it! Show me 10!**
- Say, **Let's play our game.**
- Shuffle the Number Recognition Cards and show them one at a time. Go around the group three times. Correct errors.

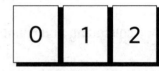

Number Recognition Cards

TEN FRAMES

MATERIALS: Teacher Ten Frame Mat with 10 two-colored dots, white board magnetic easel

★ Making 5, 6, 7, 8, 9, 10

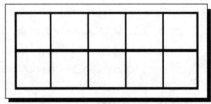

Ten Frame Mat

- Put 5 red dots in the top row and 5 yellow dots in the bottom row.
- Put the Ten Frame Mat on the white board magnetic easel and say, **Here is a Ten Frame. 5 and 5.** Point to the two sets of blocks. **5 blocks for this hand and 5 blocks for this hand.** Show each hand, one at a time. **5 and 5.**
- Say, **We know how to make 6 on our fingers. 5 and 1.** Demonstrate on fingers. **It looks the same on our frame. 5 and 1.** Put the dots on the Ten Frame.
- Say, **See, 5 and 1; 5, 6.** Draw your finger across the row of 5 when saying, "5," point to the single dot in the bottom row when saying, "1 or 6."
- Repeat for 5 + 2, 5 + 3, 5 + 4, 5 + 5.

★ Ten Frame Game

MATERIALS: Student Ten Frame Trays with 10 two-colored magnetic dots, Number Recognition Cards 5–10

- Hand out the Student Ten Frame Trays and say, **Now it is your turn! I am going to show you a number from 5 to 10, and I want you to quickly make it on your Ten Frame and show me when you are finished. Keep the top dots red and the bottom dots yellow. Ready?**

Number Sense Interventions Lesson 10   71

- Hold up a Number Recognition Card in random order and have children build it on their Ten Frames. They should hold up their Ten Frame so that it faces you when they are finished.
- If a child is incorrect, then have him or her count, but make sure he or she counts starting at 5, not 1.
- Collect the Ten Frames.

★ Recognizing Sets

MATERIALS: Ten Frame Flash Cards

- Say, **Now I am going to hold up a card with some circles on it, and I want you to tell me how many circles are on the card. Try to tell me as fast as you can without counting. I will point to you for your turn, okay? Everyone else, say it in your mind.**
- Go through the Ten Frame Flash Cards in random order. Go around the group three times. Correct errors.
- Give each child two number sentences to show.

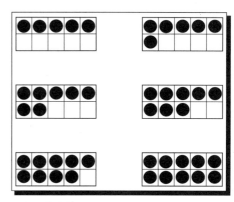
Ten Frame Flash Cards

COMPARING QUANTITIES

MATERIALS: Ten Frame Flash Cards

- Go through the Ten Frame Flash Cards saying, **Listen carefully. Which has more?** Respond with, **That is right! ___ is more than ___.** Some children may naturally start saying the numbers but do not require it.
- Show 5 pairs. Shuffle the cards.
- Say, **Now we are going to do something different. Listen carefully.**
- Go through the Ten Frame Flash Cards saying, **Listen carefully. Which has less?** Respond with, **That is right! ___ is less than ___.** Some children may naturally start saying the numbers but do not require it.
- Show 5 pairs.

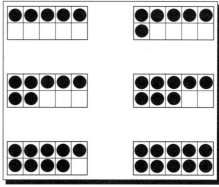
Ten Frame Flash Cards

STORY PROBLEMS

MATERIALS: Number Sentence Cards

- Say, **I am going to tell you some stories. Make the stories on your fingers.** Put the Number Sentence Cards on the table.

$5 - 1 = 4$

- Say, **Mike has 5 markers. Show me Mike's markers on your fingers. His sister took 1 marker. Take away 1 marker on your fingers. How many markers does Mike have now? Tell me without counting!** Wait for a response.

| 5 − 1 = 4 |

Number Sentence Card

- Say, **Which number sentence goes with this story: 5 markers take away 1 marker?** Call on a child to choose the correct card. If he or she chooses the incorrect card, then say, **This card says 5 take away __. Which card says 5 take away 1?**

$5 - 2 = 3$

- Say, **Christine has 5 pencils. Show me Christine's pencils using your fingers. She gave Jill 2 of her pencils. Cover 2 fingers. How many pencils does Christine have now? Tell me without counting!** Wait for a response.
- Say, **Which number sentence goes with this story: 5 pencils take away 2 pencils?** Call on a child to choose the correct card. If he or she chooses the incorrect card, then say, **This card says 5 take away __. Which card says 5 take away 2?**

$5 - 2 = 3$
Number Sentence Card

$5 - 3 = 2$

- Say, **Juan picked 5 apples. Show me Juan's apples on your fingers. 3 apples had worms. Cover 3 fingers. How many apples did not have worms? Tell me without counting!** Wait for a response.
- Say, **Which number sentence goes with this story: 5 apples take away 3 apples?** Call on a child to choose the correct card. If he or she chooses the incorrect card, then say, **This card says 5 take away __. Which card says 5 take away 3?**

$5 - 3 = 2$
Number Sentence Card

$5 - 4 = 1$

- Say, **Demiah had 5 DVDs downstairs. She took 4 DVDs upstairs. Cover up 4 fingers.** Demonstrate. **How many videos are still downstairs? Tell me without counting!** Wait for a response.
- Say, **Which number sentence goes with this story: 5 DVDs take away 4 DVDs?** Call on a child to choose the correct card. If he or she chooses the incorrect card, then say, **This card says 5 take away __. Which card says 5 take away 4?**

$5 - 4 = 1$
Number Sentence Card

NUMBER SENTENCES ON FINGERS

MATERIALS: Number Sentence Cards

- Pick up the Number Sentence Cards and say, **Now we are going to show these number sentences on our fingers.**
- Say, **I will hold up a card and you make it on your fingers. Remember, we can show 5 on one hand by showing all our fingers. Show one hand. We do not need to count!**
- Say, **Let's try this one. It says 5 plus 1. Let's show 5 plus 1.** Show one hand and 1 finger.
- Say, **I am going to give you each a number sentence to show on your fingers.**

$5 + 1 = 6$
Number Sentence Card

WRITTEN NUMBERS

MATERIALS: Lesson 10 Activity Sheets, pencils, crayons

- Say, **Now we are going to practice writing numbers. Here is a paper with the numbers 9 and 10. I would like you to trace over the numbers with your pencil. Do your best to stay on the dotted lines, but do not erase if you make a mistake. Just try it again! Under the dotted numbers is a place to copy the numbers on your own. Then write the number that tells us how many things are in the box.** Point to the pictures at the bottom of the page.
- As each child finishes, have him or her turn the paper over and say, **Write the missing number for these number sentences. If you do not know the missing number, then use your fingers to figure it out.**

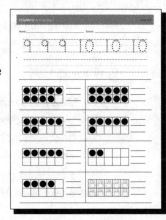
Lesson 10 Activity Sheet

Number Sense Interventions Lesson 11   73

Lesson 11

Learning Goals	Materials
Establish behavior boundaries Count to 40 orally Build 11 as 10 and 1 more Count and sequence to 11 Number recognition 0–11 Identify bigger/smaller number Identify number "after" using Number List Make numbers 1–10 on fingers Represent 5–10 on Ten Frame Recognize quantities 0–10 on Ten Frame Story problems and number sentences: $5 + n$, $5 - n$ Perform number operations on fingers Find missing numbers on Number List Solve number sentences: Mixed practice 2–5 families	**COPY** Hundreds Chart Cardinality Chart Decade Card (10) Unit Card (1) Bigger/Smaller Cards Teacher Number List Number Sentence Cards: $5 + 2 = 7$, $5 + 3 = 8$, $5 + 4 = 9$, $5 + 5 = 10$, $5 - 2 = 3$, $5 - 4 = 1$ Lesson 11 Activity Sheet **GATHER** White board magnetic easel 11 interlocking blocks Sticky notes Pencils without erasers and crayons **PREPARE** Put student names on Lesson 11 Activity Sheets. Put sticky notes on Teacher Number List Number Recognition Cards (0–11); *See Chapter 1 for instructions on making Number Recognition Cards.* Teacher Ten Frame Mat with 10 two-colored dots; *See Chapter 1 for instructions on making the Ten Frame Mat and two-colored dots.* Student Ten Frame Trays with 10 two-colored dots

ESTABLISHING BEHAVIOR BOUNDARIES

- Say, **Show me you are ready to learn.** Wait for proper posture. **Exactly! That shows me you are ready!**

COUNTING WARM-UP

MATERIALS: Hundreds Chart, white board magnetic easel

- Put up the Hundreds Chart to use for reference when needed.
- Say, **Let's count to 40 this time, but we will start counting at 11. I will start. 11** (point to the child to your right). If a child says the incorrect number, then say, **That was a good try, but the next number is 12. Let's try again.** For each error, back up 2 children and repeat so the child has an opportunity to be successful. Take a turn yourself.

Hundreds Chart

MAGIC NUMBER ACTIVITIES (Magic Number is 11)

★ Making the Number 11

MATERIALS: 10 Decade Card, 1 Unit Card, Cardinality Chart, 11 interlocking blocks

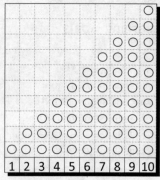
Cardinality Chart [number chart]

- Put down 11 unattached interlocking blocks and the Cardinality Chart. Say, **Let's count these blocks. We are going to put them together in a stick like we did for our chart.** Count the blocks, making a stick until you reach 10.
- Put the stick on the Cardinality Chart above 10. Say, **How many blocks are here?** Wait for a response.
- Say, **That is right. Just like 10 fingers. Watch this!** Take the stick off of the chart. Put 1 finger (starting with your pinky) on each block as you count to 10 again. Say, **10 blocks, 10 fingers!**
- Say, **Each time we count to 10, we will make a stick of 10 to remember we used all our fingers! Now we can keep counting.**
- Run your finger up the stick of 10 as you say, "10." Point to the 1 block as you say, "11." Repeat with gestures.
- Circle all the blocks and say, **How many blocks do we have altogether?** Wait for a response.
- Say, **That is right. There are 11 blocks altogether.**
- Hold up the 10 Decade Card and say, **What number is this?** Wait for a response. Say, **Yes, the number is 10.**
- Say, **I will put the number 10 right next to the 10 blocks.** Lay the stick down and put down the 10 Decade Card next to the stick.
- Say, **What number is this?** Wait for a response. **That is right. This is the number 1. I will put the 1 Unit Card right next to the 1 block.**
- The table should look like the following:

10
Decade Card

1
Unit Card

Number Sense Interventions Lesson 11   75

- Say, **Now watch what I do.** Put the 1 Unit Card over the 0 on the Decade Card while saying, **10 and 1 more is 11. This is the number 11. What number is this?** Wait for a response.
- Say, **That is right. This is the number 11. See the 10 hiding under the 1?** Lift up the 1 Unit Card. **11 is 10** (show the 10) **and 1 more.** Add the 1 on top.
- Then say, **10** (run your finger up the stick of 10) **and 1 more** (point to the one block) **is 11.** Repeat.

★ Sequencing and Number Recognition 0–11

MATERIALS: Number Recognition Cards 0–11

Number Recognition Cards

- Say, **Say the numbers with me as I put them down. Do not go ahead of me. Ready? 1, 2, 3 … 11!** Put the cards down horizontally as if building a number list. Orient the cards toward the children.
- Hold up the Number Recognition Card 11 and say, **What number is this?** Wait for a response.
- Say, **That is right. This is the number 11. Our Magic Number today is 11.**
- Say, **We are going to put the 11 under the 1 card. 11 is 10** (run your finger along the number list 1–10), **and 1 more** (point to the 11 Number Recognition Card). Repeat.
- Say, **Let's play our game!**
- Go around the group, showing each child a different number from 0 to 11 (not in order). Make sure all children can see the number. Go around the group three times.

BIGGER/SMALLER

MATERIALS: Cardinality Chart, Bigger/Smaller Cards, white board magnetic easel

- Put the Cardinality Chart on the white board magnetic easel. Say, **As I go up the number list, each number has 1 more circle than the one before it. See how there are more circles as I go up the number list. When I say a number is bigger, I mean it has more circles.**
- Say, **Which number is bigger or more: 4 or 9? Which has more circles?** Point to the numbers as you say them. Wait for a response. **That is right. 9 is bigger than 4.** If they make an error, then point to the numbers on the Cardinality Chart and say, "Which has more circles?"
- Say, **Let's try another one. Which is bigger or more: 8 or 6? Which has more circles?** Point to the numbers as you say them. Wait for a response. **That is right. 8 is bigger than 6.**
- Say, **I will show you a card with 2 numbers on it, and you tell me which number is bigger.**
- Go around twice, showing the Bigger/Smaller Cards in random order, saying, **Which number is bigger or more?**
- Say, **Now we are going to find smaller numbers. I will show you a card with 2 numbers on it, and you tell me which number is smaller—which number has less circles. Let's try one together. Which is smaller: 4 or 6?** Wait for a response.
- Say, **That is right. 4 is smaller than 6. Now it is your turn.**
- Go around the group twice, showing the Bigger/Smaller Cards in random order, saying, **Which number is smaller or less?** Use the same error correction.

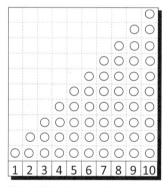
Cardinality Chart [number chart]

Bigger/Smaller Cards

NUMBER LIST—NUMBER AFTER

MATERIALS: Teacher Number List, sticky notes

- Take out the Teacher Number List and say, **This is a number list. It has the numbers 1-10 in order, just like our Cardinality Chart. But the numbers are covered up!**

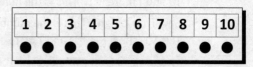
Teacher Number List

- Say, **What is the first number we say when we count?** Wait for a response. **So what number do you think is hiding here** (point to the number 1 space)?
- Say, **Let's see if we are right!** Remove the sticky note over 1 and say, **That is right. Our first counting number is 1.**
- Say, **The number right after 1 is the next number up** (gesture up the Teacher Number List) **the number list** (point to the 2 space). **What number comes right after 1? What number do you think is hiding here?** Wait for a response.
- While removing the sticky note say, **That is right. The number is 2. The number right after 1 is 2.**
- Say, **The number right after 2 is the next number up** (gesture up the Teacher Number List) **the number list** (point to the 3 space). **What number comes after 2? What number do you think is hiding here?** Wait for a response.
- While removing the sticky note say, **That is right. The number is 3. The number right after 2 is 3.**
- Repeat until you reach 10. If the students are behaving, then allow them to take off the sticky notes.

SUBITIZING QUANTITIES (1-10) ACTIVITIES

★ Finger Automaticity

MATERIALS: Number Recognition Cards 1-10

- Say, **Show me 1.** Repeat with numbers 2-10.
- Say, **Let's play our game.**
- Shuffle the Number Recognition Cards and show them one at a time. Go around the group three times. Correct errors.

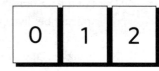
Number Recognition Cards

TEN FRAMES

MATERIALS: Teacher Ten Frame Mat with 10 two-colored dots, white board magnetic easel

★ Making 5, 6, 7, 8, 9, 10

- Put 5 red dots in the top row and 5 yellow dots in the bottom row.
- Put the Ten Frame Mat on the white board magnetic easel and say, **Here is a Ten Frame. 5 and 5.** Point to the 2 sets of blocks. **5 blocks for this hand and 5 blocks for this hand.** Show each hand, one at a time. **5 and 5.**

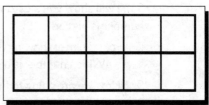
Ten Frame Mat

- Say, **We know how to make 6 on our fingers. 5 and 1.** Demonstrate on fingers. **It looks the same on our frame. 5 and 1.** Put the dots on the Ten Frame Mat.
- Say, **See, 5 and 1; 5, 6.** Draw your finger across the row of 5 when saying, "5," point to the single dot in the bottom row when saying, "1 or 6."
- Repeat for 5 + 2, 5 + 3, 5 + 4, 5 + 5.

★ Ten Frame Game

MATERIALS: Student Ten Frame Trays with 10 two-colored magnetic dots, Number Recognition Cards 5-10

- Hand out the Ten Frame Trays and say, **Now it is your turn! I am going to show you a number from 5 to 10, and I want you to quickly make it on your Ten Frame and show me when you are finished. Keep the top dots red and the bottom dots yellow. Ready?**
- Hold up a Number Recognition Card in random order and have children build it on their Ten Frames. They should hold up their Ten Frame so that it faces you when they are finished.
- If a child is incorrect, then have him or her count, but make sure he or she counts starting at 5, not 1.
- Collect the Ten Frames.

Number Recognition Cards

STORY PROBLEMS

MATERIALS: Number Sentence Cards

- Say, **I am going to tell you some stories. Make the stories on your fingers.** Put the Number Sentence Cards on the table.

5 + 2 = 7

- Say, **Mike has 5 pennies. Show me Mike's pennies on your fingers. His sister gave him 2 more pennies. How many pennies does Mike have now? Tell me without counting!** Wait for a response.
- Say, **Which number sentence goes with this story: 5 pennies and 2 pennies more?** Call on a child to choose the correct card.

5 + 4 = 9

- Say, **Christine has 5 stickers. Show me Christine's stickers on your fingers. Jill gave her 4 of her stickers. How many stickers does Christine have now? Tell me without counting!** Wait for a response.
- Say, **Which number sentence goes with this story: 5 stickers and 4 more stickers?** Call on a child to choose the correct card.

5 − 2 = 3

- Say, **This problem is a little different. Juan picked 5 apples. Show me Juan's apples on your fingers. 2 of the apples had worms. Cover 2 fingers. How many apples did not have worms? Tell me without counting!** Wait for a response.
- Say, **Which number sentence goes with this story: 5 apples take away 2 apples?** Call on a child to choose the correct card.

5 − 4 = 1

- Say, **Demiah had 5 stickers. She lost 4. Cover 4 fingers. How many stickers does she still have? Tell me without counting!** Wait for a response.
- Say, **Which number sentence goes with this story: 5 stickers take away 4 stickers?** Call on a child to choose the correct card.

WRITTEN NUMBERS

MATERIALS: Lesson 11 Activity Sheet, pencils, crayons

- Say, **Here are some number lists with numbers missing. It is your job to write in the missing numbers. Take your time and do your best.** Have children self-correct their papers by orally counting up the number list.

- As each child finishes, have him or her turn the paper over and say, **Write the missing number for these number sentences. If you do not know the missing number, then use your fingers to figure it out.**

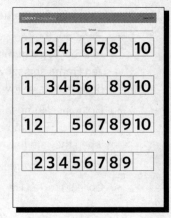

Lesson 11 Activity Sheet

Lesson 12

Learning Goals	Materials
Establish behavior boundaries	**COPY**
Count to 40 orally	Hundreds Chart
Build 12 as 10 and 2 ones	Decade Card (10)
Count and sequence to 12	Unit Card (2)
Number recognition 0–12	Teacher Number List
Before and after (1–10)	Ten Frame Flash Cards
Plus 1 using Number List	Number Sentence Cards: 5 + 3 = 8, 5 + 5 = 10, 5 – 3 = 2, 5 – 5 = 0
Make numbers 1–10 on fingers	Lesson 12 Activity Sheet
Represent 5–10 on Ten Frames	**GATHER**
Recognize quantities 0–10 on Ten Frames	White board magnetic easel
Story problems and number sentences: 5 + n, 5 – n	12 interlocking blocks
Find missing numbers on number list	Sticky notes
Solve number sentences: Mixed practice 2–5 families	10 black dots
	Pencils without erasers and crayons
	PREPARE
	Put student names on Lesson 12 Activity Sheets.
	Put sticky notes on Teacher Number List
	Number Recognition Cards (0–12); *See Chapter 1 for instructions on making Number Recognition Cards.*
	Teacher Ten Frame Mat with 10 two-colored dots; *See Chapter 1 for instructions on making the Ten Frame Trays with two-colored dots.*
	Student Ten Frame Trays with 10 two-colored dots; *See Chapter 1 for instructions on making the Ten Frame Trays with two-colored dots.*

ESTABLISHING BEHAVIOR BOUNDARIES

- Say, **Show me you are ready to learn.** Wait for proper posture. **Exactly! That shows me you are ready!**

COUNTING WARM-UP

MATERIALS: Hundreds Chart, white board magnetic easel

- Put up the Hundreds Chart to use for reference when needed.
- Say, **Let's count to 40 this time, but we will start counting at 11. I will start. 11** (point to the child to your right). If a child says the incorrect number, then say, **That was a good try, but the next number is 12. Let's try again.** For each error, back up 2 children and repeat so the child has an opportunity to be successful. Take a turn yourself.

1	2	3	4	5	6	7	8	9	10
11	12	13	14	15	16	17	18	19	20
21	22	23	24	25	26	27	28	29	30
31	32	33	34	35	36	37	38	39	40
41	42	43	44	45	46	47	48	49	50
51	52	53	54	55	56	57	58	59	60
61	62	63	64	65	66	67	68	69	70
71	72	73	74	75	76	77	78	79	80
81	82	83	84	85	86	87	88	89	90
91	92	93	94	95	96	97	98	99	100

Hundreds Chart

MAGIC NUMBER ACTIVITIES (Magic Number is 12)

★ Making the Number 12

MATERIALS: 10 Decade Card, 2 Unit Card, 12 interlocking blocks

- Put down 12 unattached interlocking cubes. Say, **Here are some blocks. Let's count them together. You count on your fingers while I touch the blocks. Remember to use your inside voices. No shouting. Put your hands up so I can see them.**
- Count to 10, making a stick of 10 as you count. The children will have run out of fingers at 10. Hold up your fingers and say, **10 fingers! 10 blocks** (hold up the stick of 10 blocks).
- Say, **Let's keep counting.** Hold up the stick of 10 and as you say, "10." Touch each single block as you say, "11, 12."
- Say, **Let's show that on our fingers. 10** (hold up 10 fingers then put down hands and begin counting on fingers again), **11, 12.** Make sure children are counting correctly.
- Circle all the blocks and say, **How many blocks do we have altogether?** Wait for a response. **That is right. There are 12 blocks altogether.**
- Hold up the 10 Decade Card and say, **What number is this?** Wait for a response. Say, **Yes, the number is 10.**
- Say, **I will put the number 10 right next to the 10 blocks.** Lay the stick down and put down the 10 Decade Card next to the stick.
- Hold up the 2 Unit Card and say, **What number is this?** Wait for a response. **That is right. This is the number 2. I will put the number 2 right next to the two blocks.**

10

Decade Card

2

Unit Card

- The table should look like the following.

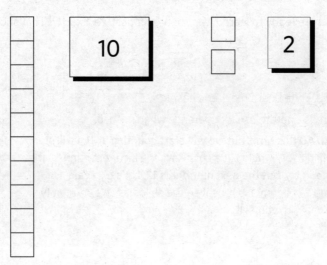

- Say, **Now watch what I do**. Put the 2 Unit Card over the 0 on the Decade Card while saying, **10 and 2 more is 12. This is the number 12. What number is this?** Wait for a response.
- Say, **That is right. This is the number 12. See the 10 hiding under the 2** (lift up the 2 Unit Card)? **12 is 10** (show the Decade Card) **and 2 more** (add the 2 Unit Card on top).
- Then say, **10** (run your finger up the stick of 10) **and 2 more is 11, 12** (touch each single block as you say, "11, 12"). Repeat.
- Then say, **10** (run your finger up the stick of 10) **and 2 more** (point to the 2 blocks) **is 12**. Repeat.

★ Sequencing and Number Recognition 0–12

MATERIALS: Number Recognition Cards 0–12

- Say, **Say the numbers with me as I put them down. Do not go ahead of me. Ready? 1, 2, 3 … 12!** Put the cards in 2 rows as if building a Hundreds Chart.
- Hold up the Number Recognition Card 12 and say, **What number is this?** Wait for a response.
- Say, **That is right. This is the number 12. Our Magic Number today is 12.**
- Say, **We are going to put the 12 here, under the 2 card. 12 is 10** (run your finger along the number list 1–10), **and 2 more** (point to the 11 and 12 Number Recognition Cards). Repeat.
- Say, **Let's play our game!**
- Go around the group, showing each child a different number from 0 to 12 (not in order). Make sure all children can see the number. Go around the group three times.

★ Before and After

MATERIALS: Number Recognition Cards 1–10

Number Recognition Cards

- Say, **I am going to show you a number, and I want you to tell me what number comes right after it. We will take turns, so do not call out if it is not your turn. Think carefully and tell me what number comes after.**
- Show Number Recognition Cards 1–10 in random order. As you hold up a card say, **What number comes right after _____?**
- Error correction: If the child gives a wrong answer, then put the card on the table and say, **What number comes after/before _____** and point to the space after as in previous activities.
- Give each child two turns.
- Repeat with before numbers.

NUMBER LIST—Number +1

MATERIALS: Teacher Number List, black dots, sticky notes

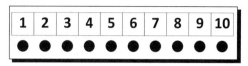
Teacher Number List

- Take out the Teacher Number List and 10 dots and say, **This is a number list. We begin at the dot under number 1 when we count using the number list.** Say, **Today we are going to count these dots using the number list.**

- Say, **I will put 1 dot on the number list. We always start counting with the dot under number 1. See, there is 1 dot. This is the number 1.** Point to the number 1. **If I put on 1 more dot, then how many dots will I have on the number list?** Wait for a response. While putting 1 more dot on the Teacher Number List say, **That is right. 1 and 1 more makes 2** (circle the 1 dot when you say, "1" and the 2 dots when you say, "2").

- Say, **What number do you think is hiding here?** Point to the sticky note over 2. Wait for a response. While removing the sticky note say, **That is right. The number is 2 because that tells you how many there are in all. 1 and 1 more makes 2** (circle the 2 dots).

- **If I put on 1 more dot, then how many dots will I have on the number list?** Wait for a response. While putting 1 more dot on the number list say, **That is right. 2 and 1 more makes 3** (circle the 2 dots when you say, "2" and the 3 dots when you say, "3").

- Say, **What number do you think is hiding here?** Point to the sticky note over 3. Wait for a response. While removing the sticky note say, **That is right. The number is 3 because that tells you how many there are in all. 2 and 1 more makes 3** (circle 3 dots).

- Repeat until you reach 10. If the students are behaving, then allow them to put the dots on the list and take off the sticky notes. Put away the dots.

SUBITIZING QUANTITIES (1–10) ACTIVITIES

★ Finger Automaticity

MATERIALS: Number Recognition Cards 1–10

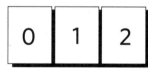
Number Recognition Cards

- Say, **Let's play our finger game. I will show you a card, and you show me the number on your fingers. Be as quick as you can!**

- Shuffle the Number Recognition Cards and show them one at a time. Go around the group three times. Correct errors.

★ Ten Frame Game

MATERIALS: Student Ten Frame Tray with 10 two-colored magnetic dots, Number Recognition Cards 5–10

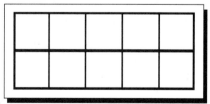
Ten Frame Mat

- Hand out the Student Ten Frame Trays and say, **Now it is your turn! I am going to show you a number from 5 to 10, and I want you to quickly make it on your Ten Frame and show me when you are finished. Keep the top dots red and the bottom dots yellow. Ready?**

- Hold up a Number Recognition Card in random order and have children build it on their Ten Frames. They should hold up their Ten Frame so that it faces you when they are finished.

- If a child is incorrect, then have him or her count, but make sure he or she counts starting at 5, not 1.

- Collect the Ten Frames.

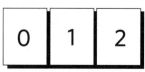
Number Recognition Cards

★ Recognizing Sets

MATERIALS: Ten Frame Flash Cards

- Say, **Now I am going to hold up a card with some circles on it, and I want you to tell me how many circles are on the card. Try to tell me as fast as you can without counting. I will point to you for your turn, okay? Everyone else, say it in your mind.**
- Go through the Ten Frame Flash Cards in random order. Go around the group three times. Correct errors.

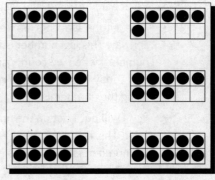
Ten Frame Flash Cards

STORY PROBLEMS

MATERIALS: Teacher Ten Frame Mat with 10 two-colored dots, student Ten Frame with 10 two-colored dots, Number Sentence Cards

- Hand out Student Ten Frame Trays. Say, **I am going to tell you some stories. Make the stories on your frames.**

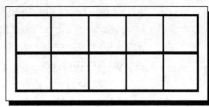
Ten Frame Mat

5 + 3 = 8

- Say, **Mike has 5 markers. Show me Mike's markers using red dots. His sister gave him 3 more markers. Use yellow dots to show 3 more. How many markers does Mike have now? Tell me without counting?** Wait for a response.
- Say, **Which number sentence goes with this story: 5 markers plus 3 markers?** Call on a child to choose the correct card.

5 + 3 = 8
Number Sentence Card

5 + 5 = 10

- Say, **Christine has 5 pencils. Show me Christine's pencils using your dots. Jill gave her 5 more pencils. How many pencils does Christine have now? Tell me without counting.** Wait for a response.
- Say, **Which number sentence goes with this story: 5 pencils plus 5 pencils?** Call on a child to choose the correct card.

5 + 5 = 10
Number Sentence Card

5 − 3 = 2

- Say, **Juan picked 5 apples. Show me Juan's apples using your dots. 3 apples had worms.** Turn over 3 dots on the Teacher Ten Frame for the 3 worms. **How many apples did not have worms? Tell me without counting.** Wait for a response.
- Say, **Which number sentence goes with this story: 5 apples take away 3 apples?** Call on a child to choose the correct card.

5 − 3 = 2
Number Sentence Card

5 − 5 = 0

- Say, **Demiah had 5 stickers. She lost 5.** Turn over 5 dots on the Teacher Ten Frame Mat. **How many stickers does she still have? Tell me without counting.** Wait for a response.
- Say, **Which number sentence goes with this story: 5 stickers take away 5 stickers?** Call on a child to choose the correct card.

5 − 5 = 0
Number Sentence Card

Number Sense Interventions Lesson 13

WRITTEN NUMBERS

MATERIALS: Lesson 12 Activity Sheets, pencils, crayons

- Say, **Here are some number lists, but some of the numbers are missing. I would like you to fill in the missing numbers. When you are finished, turn the paper over and finish the number sentences.**
- Hand out the Lesson 12 Activity Sheet and pencils. If a child finishes early, then check his or her work. If it is incorrect, then circle the error and have him or her correct it. If it is correct, then allow him or her to color the number list.

Lesson 12 Activity Sheet

Lesson 13

Learning Goals	Materials
Establish behavior boundaries	**COPY**
Count to 50 orally	Hundreds Chart
Build 13 as 10 and 3 ones	Cardinality Chart
Count and sequence to 13	Decade Card (10)
Number recognition 0-13	Unit Card (3)
Make numbers 1-10 on fingers	Bigger/Smaller Cards
Identify bigger/smaller number	Teacher Number List
Number "before" using Teacher Number List	Ten Frame Flash Cards
Represent 5-10 on Ten Frames	Lesson 13 Activity Sheets
Recognize quantities 0-10 on Ten Frames	**GATHER**
Number Sentences: 5 + n, 5 − n	White board magnetic easel
Perform number operations on fingers	13 interlocking blocks
Find "before" and "after" numbers	Sticky notes
Solve number sentences: Mixed practice 2-5 families	Pencils without erasers and crayons
	PREPARE
	Put student names on Lesson 13 Activity Sheets.
	Put sticky notes on Teacher Number List
	Number Recognition Cards (0-13); *See Chapter 1 for instructions on making Number Recognition Cards.*
	Teacher Ten Frame Mat with 10 two-colored dots; *See Chapter 1 for instructions on making the Ten Frame Mat and two-colored dots.*
	Student Ten Frame Trays with 10 two-colored dots

ESTABLISHING BEHAVIOR BOUNDARIES

- Say, **Show me you are ready to learn.** Wait for proper posture. **Exactly! That shows me you are ready!**

COUNTING WARM-UP

MATERIALS: Hundreds Chart, white board magnetic easel

- Put up the Hundreds Chart to use for reference when needed.
- Say, **Let's count to 50 this time, but we will start counting with 21. I will start. 21** (point to the child to your right). If a child says the incorrect number, then say, **That was a good try, but the next number is _____. Let's try again.** For each error, back up 2 children and repeat so the child has an opportunity to be successful. Take a turn yourself.

Hundreds Chart

MAGIC NUMBER ACTIVITIES (Magic Number is 13)

★ Making the Number 13

MATERIALS: 10 Decade Card, 3 Unit Card, 13 interlocking blocks

- Put down 13 unattached interlocking blocks. Say, **Here are some blocks. Let's count them together. You count on your fingers while I touch the blocks. Remember to use your inside voices. No shouting. Put your hands up so I can see them.**
- Count to 10, making a stick of 10 as you count. The children will have run out of fingers at 10. Hold up your fingers and say, **10 fingers! 10 blocks** (hold up the stick of 10 blocks).
- Say, **Let's keep counting.** Hold up the stick of 10 when you say, "10." Touch each single block as you say, "11, 12, 13."
- Say, **Let's show that on our fingers. 10** (hold up 10 fingers then put down hands and begin counting on fingers again), **11, 12, 13.** Make sure children are counting correctly.
- Circle all the blocks and say, **How many blocks do we have altogether?** Wait for a response. **That is right. There are 13 blocks altogether.**
- Hold up the 10 Decade Card and say, **What number is this?** Wait for a response. Say, **Yes, the number is 10.**
- Say, **I will put the number 10 right next to the 10 blocks.** Lay the stick down and put down the 10 Decade Card next to the stick.
- Hold up the 3 Unit Card and say, **What number is this? That is right. This is the number 3. I will put the number 3 right next to the 3 blocks.**

Decade Card

Unit Card

Number Sense Interventions Lesson 13   85

- The table should look like the following.

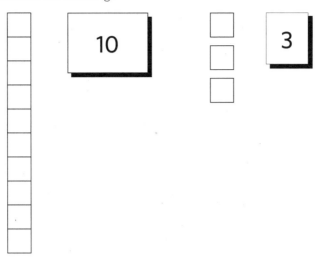

- Say, **Now watch what I do.** Put the 3 Unit Card over the 0 on the 10 Decade Card while saying, **10 and 3 more is 13. This is the number 13. What number is this?** Wait for a response.
- Say, **That is right. This is the number 13.** See the 10 hiding under the 3 (lift up the 3 Unit Card)? **13 is 10** (show the 10) **and 3 more** (add the 3 Unit Card on top).
- Then say, **10** (run your finger up the stick of 10) **and 3 more is 11, 12, 13** (touch each single block as you say, "11, 12, 13"). Repeat.
- Then say, **10** (run your finger up the stick of 10) **and 3 more** (point to the 3 blocks) **is 13.** Repeat.

★ Sequencing and Number Recognition 0-13

MATERIALS: Number Recognition Cards 0-13

- Say, **Say the numbers with me as I put them down. Do not go ahead of me. Ready? 1, 2, 3 ... 13!** Put the cards in 2 rows as if building a Hundreds Chart.
- Hold up the Number Recognition Card 13 and say, **What number is this?** Wait for a response.
- Say, **That is right. This is the number 13. Our Magic Number today is 13.**
- Say, **We are going to put the 13 here, under the 3 card. 13 is 10** (run your finger along the number list 1-10) **and 3 more** (point to the Number Recognition Card 13). Repeat.
- Say, **Let's play our game!**
- Go around the group, showing each child a different number from 0 to 13 (not in order). Make sure all children can see the number. Go around the group three times.

| 0 | 1 | 2 |

Number Recognition Cards

BIGGER/SMALLER

MATERIALS: Cardinality chart, Bigger/Smaller Cards, white board magnetic easel

- Put the Cardinality Chart on the white board magnetic easel. Say, **Remember, as I go up the number list, each number has 1 more circle than the one before it. See how there are more circles as I go up the number list. When I say a number is bigger, I mean it has more circles.**
- Say, **Which number is bigger or more: 3 or 7? Which has more circles?** Point to the numbers as you say them. Wait for a response. **That is right. 7 is bigger than 3.** If they make an error, then point to the numbers on the Cardinality Chart and say, "Which has more dots?"

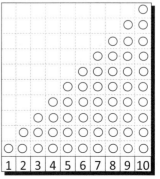

Cardinality Chart [number chart]

- Say, **Let's try another one. Which is bigger or more: 9 or 6? Which has more circles?** Point to the numbers as you say them. Wait for a response. **That is right. 9 is bigger than 6.**
- Say, **I will show you a card with 2 numbers on it, and you tell me which number is bigger.**
- Go around the group twice and show the Bigger/Smaller Cards in random order saying, **Which number is bigger or more?**
- Say, **Now we are going to find smaller numbers. I will show you a card with 2 numbers on it, and you tell me which number is smaller—which one has less circles. Let's try one together. Which is smaller: 3 or 5? Which has less circles?** Wait for a response.
- Say, **That is right. 3 is smaller than 5. Now it is your turn.**
- Go around the group twice and show the Bigger/Smaller Cards in random order saying, **Which number is smaller or less?** Use the same error correction.

Bigger/Smaller Cards

NUMBER LIST—NUMBER BEFORE

MATERIALS: Teacher Number List, sticky notes

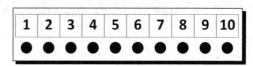

Teacher Number List

- Take out the Teacher Number List and say, **Here is my number list. Remember, it has the numbers from 1-10 in order, just like our Cardinality Chart. These are the numbers we use to count. But the numbers are covered up!**
- Say, **This time we are going to go down** (gesture down) **the number list and say the numbers before and not the numbers after.** Point to the 10 space and say, **What number do you think is hiding here?** Wait for a response.
- Say, **Let's see if we are right!** Remove the sticky note over 10 and say, **That is right. Our last number on the list is 10.**
- Say, **The number right before 10 is the next number down** (gesture down) **the number list.** Point to the 9 space. **What number comes right before 10? What number do you think is hiding here?** Wait for a response.
- If the children respond incorrectly, then touch the space on the number list beginning with 1 and count up to the space before 10.
- While removing the sticky note say, **That is right. The number is 9. The number right before 10 is 9.**
- **The number right before 9 is the next number down** (gesture down) **the number list.** Point to the 8 space. **What number comes right before 9? What number do you think is hiding here?** Wait for a response.
- Repeat error correction if necessary.
- While removing the sticky note say, **That is right. The number is 8. The number right before 9 is 8.**
- Repeat until you reach 1. If the students are behaving, then allow them to take off the sticky notes.
- Say, **Remember, when we find numbers before, we count down the list, just like when you do a count down! Let's count down from 10.** Touch the numbers on the Teacher Number List as you count down. **10, 9, 8 … 1!**

SUBITIZING QUANTITIES (1-10) ACTIVITIES

★ Finger Automaticity

MATERIALS: Number Recognition Cards 1-10

- Say, **Let's play our finger game. I will show you a card, and you show me the number on your fingers. Be as quick as you can!**
- Shuffle the Number Recognition Cards and show them one at a time. Go around the group three times. Correct errors.

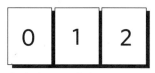

Number Recognition Cards

★ Making 5, 6, 7, 8, 9, 10

MATERIALS: Teacher Ten Frame Mat and 10 two-colored dots, white board magnetic easel

- Put the Ten Frame Mat on the white board magnetic easel and say, **Here is a Ten Frame. 5 and 5.** Point to the 2 sets of blocks. **5 blocks for this hand and 5 blocks for this hand.** Show each hand, one at a time. **5 and 5.**
- Say, **We know how to make 6 on our fingers. 5 and 1.** Demonstrate on fingers. **It looks the same on our frame. 5 and 1.** Put the dots on the Ten Frame Mat.
- Say, **See, 5 and 1; 5, 6.** Draw your finger across the row of 5 when saying, "5," point to the single dot in the bottom row when saying, "1 or 6."
- Repeat for 5 + 2, 5 + 3, 5 + 4, 5 + 5.

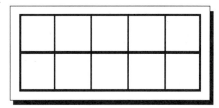

Ten Frame Mat

★ Recognizing Sets

MATERIALS: Ten Frame Flash Cards

- Say, **Now I am going to hold up a card with some circles on it, and I want you to tell me how many circles are on the card. Try to tell me as fast as you can without counting. I will point to you for your turn, okay? Everyone else, say it in your mind.**
- Go through the Ten Frame Flash Cards in random order. Go around the group three times. Correct errors.

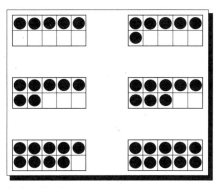

Ten Frame Flash Cards

WRITTEN NUMBERS

MATERIALS: Lesson 13 Activity Sheets, pencils, crayons

- Say, **Here are parts of some number lists, but some of the numbers are missing. I would like you to fill in the missing numbers. When you are finished, turn the paper over and finish the number sentences.**
- Hand out the Lesson 13 Activity Sheets and pencils. If a child finishes early, then check his or her work. If it is incorrect, then circle the error and have him or her correct it. If it is correct, then allow him or her to color the number list.

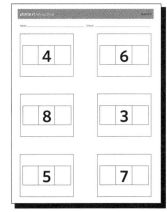

Lesson 13 Activity Sheet

Lesson 14

Learning Goals	Materials
Establish behavior boundaries Count to 80 orally and by tens to 100 Build 14 as 10 and 4 ones Count and sequence to 14 Number recognition 0–14 Before and after (1–10) Minus 1 using Teacher Number List Make numbers 1–10 on fingers Perform number operations on fingers Find "before" and "after" numbers Solve number sentences: Vertical problems	**COPY** Hundreds Chart Decade Card (10) Unit Card (4) Teacher Number List Ten Frame Flash Cards Lesson 14 Activity Sheet **GATHER** White board magnetic easel 14 interlocking blocks Sticky notes Black dots Pencils without erasers and crayons **PREPARE** Put student names on Lesson 14 Activity Sheets. Put sticky notes on Teacher Number List Number Recognition Cards (0–14); *See Chapter 1 for instructions on making Number Recognition Cards.*

ESTABLISHING BEHAVIOR BOUNDARIES

- Say, **Show me you are ready to learn.** Wait for proper posture. **Exactly! That shows me you are ready!**

COUNTING WARM-UP

MATERIALS: Hundreds Chart, white board magnetic easel

- Put up the Hundreds Chart on the white board magnetic easel.
- Say, **Last time we counted to 50. Today we are going to count even higher!**
- Gesture to the Hundreds Chart and say, **See how the numbers are lined up. Each row has 10 blocks. We can count by tens down this side.** Point to the far right column.
- Say, **Let's count by tens together. 10, 20, 30, 40, 50. Let's keep going. 60, 70, 80, 90, 100.** Draw your finger across each row and end on the decade as you count.
- Say, **We can count higher now that we know the pattern! Let's start at 45. I will start. 45.** Point to the child to your right. Use the Hundreds Chart for reference and error correction. Take a turn yourself.
- Continue to count to 80.

Hundreds Chart

Number Sense Interventions Lesson 14   89

MAGIC NUMBER ACTIVITIES (Magic Number is 14)

★ Making the Number 14

MATERIALS: 10 Decade Card, 4 Unit Card, 14 interlocking blocks

- Put down 14 unattached interlocking blocks. Say, **Here are some blocks. Let's count them together. You count on your fingers while I touch the blocks. Remember to use your inside voices. No shouting. Put your hands up so I can see them.**
- Count to 10, making a stick of 10 as you count. The children will have run out of fingers at 10. Hold up your fingers and say, **10 fingers! 10 blocks** (hold up the stick of 10 blocks).
- Say, **Let's keep counting.** Hold up the stick of 10 and say, "10." Touch each single block as you say, "11, 12, 13, 14."
- Say, **Let's show that on our fingers. 10** (hold up 10 fingers then put down hands and begin counting on fingers again), **11, 12, 13, 14.** Make sure children are counting correctly.
- Circle all the blocks and say, **How many blocks do we have altogether?** Wait for a response. **That is right. There are 14 blocks altogether.**
- Hold up the 10 Decade Card and say, **What number is this?** Wait for a response. Say, **Yes, the number is 10.**
- Say, **I will put the number 10 right next to the 10 blocks.** Lay the stick down and put down the 10 Decade Card next to the stick.
- Hold up the 4 Unit Card and say, **What number is this?** Wait for a response. **That is right. This is the number 4. I will put the number 4 right next to the 4 blocks.**
- The table should look like the following.

Decade Card

Unit Card

- Say, **Now watch what I do.** Put the 4 Unit Card over the 0 on the Decade Card while saying, **10 and 4 more is 14. This is the number 14. What number is this?** Wait for a response.
- Say, **That is right. This is the number 14. See the 10 hiding under the 4** (lift up the 4 Unit Card)? **14 is 10** (show the 10) **and 4 more** (add the 4 Unit Card on top).
- Then say, **10** (run your finger up the stick of 10) **and 4 more is 11, 12, 13, 14** (touch each single block as you say, "11, 12, 13, 14"). Repeat.
- Then say, **10** (run your finger up the stick of 10) **and 4 more** (point to the 4 blocks) **is 14.** Repeat.

★ Sequencing and Number Recognition 0-14

MATERIALS: Number Recognition Cards 0-14

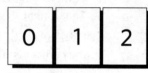
Number Recognition Cards

- Say, **Say the numbers with me as I put them down. Do not go ahead of me. Ready? 1, 2, 3 … 14!** Put the cards in two rows as if building a Hundreds Chart.
- Hold up the Number Recognition Card 14 and say, **What number is this?** Wait for a response.
- Say, **That is right. This is the number 14. Our Magic Number today is 14.**
- Say, **We are going to put the 14 here, under the 4. 14 is 10** (run your finger along the numbers 1-10) **and 4 more** (point to the Number Recognition Card 14). Repeat.
- Say, **Let's play our game!**
- Go around the group, showing each child a different number from 0 to 14 (not in order). Make sure all children can see the number. Go around the group three times.

★ Before and After

MATERIALS: Number Recognition Cards 1-10

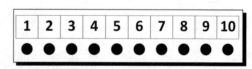
Number Recognition Cards

- Say, **I am going to show you a number, and I want you to tell me what number comes right after it. We will take turns, so do not call out if it is not your turn. Think carefully and tell me what number comes after.**
- Show the Number Recognition Cards (1-10) in random order. As you hold up a card say, **What number comes right after _____?**
- Error correction: If the child gives a wrong answer, then put the card on the table and say, **What number comes after/before _____**, and point to the space after as in previous activities.
- Give each child 2 turns.
- Repeat with "before" numbers.

NUMBER LIST—NUMBER −1

MATERIALS: Teacher Number List, black dots, sticky notes

- Take out Teacher Number List with sticky notes and 10 black dots and say, **Here is our number list. We begin at the dot under the number 1 when we count using the number list.** Say, **Let's count these dots using the number list.**
- Put the 10 dots on the Teacher Number List while counting. Circle the dots and say, **How many dots are there?** Wait for a response.
- Say, **If I take away one dot** (take away one dot), **then how many do I have now?** Wait for a response.
- Say, **What number do you think is hiding here?** Point to the sticky note over the 9. Wait for a response. While removing the sticky note say, **That is right. The number is 9 because that tells you how many are left. 10 take away 1 makes 9** (circle the 9 dots).
- Repeat until you reach 0. If the students are behaving, then allow them to take off the dots on the list and take off the sticky notes.

Number Sense Interventions Lesson 14   91

SUBITIZING QUANTITIES (1–10) ACTIVITIES

★ Finger Automaticity

MATERIALS: Number Recognition Cards 1–10

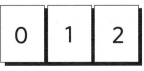
Number Recognition Cards

- Say, **Let's play our finger game. I will show you a card, and you show me the number on your fingers. Be as quick as you can!**
- Shuffle the Number Recognition Cards and show them one at a time. Go around the group three times. Correct errors.

★ Recognizing Sets

MATERIALS: Ten Frame Flash cards

- Say, **Now I am going to hold up a card with some circles on it, and I want you to tell me how many circles are on the card. Try to tell me as fast as you can without counting. I will point to you for your turn, okay? Everyone else, say it in your mind.**
- Go through the Ten Frame Flash Cards in random order. Go around the group three times. Correct errors.

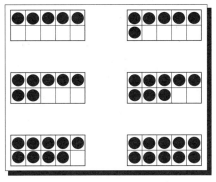
Ten Frame Flash Cards

WRITTEN NUMBERS

MATERIALS: White board magnetic easel, Lesson 14 Activity Sheets, pencils, crayons

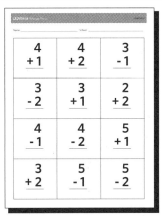
Lesson 14 Activity Sheet

- Put a copy of the Lesson 14 Activity Sheet on the white board magnetic easel.
- Say, **Sometimes number sentences are written this way, up and down. This says 4 plus 1. 4 plus 1 equals how much?** Wait for a response. If there are errors, then show on fingers—4 fingers, then add a thumb to make 5 fingers.
- Say, **Let's do this one. 4 plus 2 equals how much?**
- Say, **We can use our fingers to add 2 more. 4** (circle the 4 on the paper), **5, 6** (put up 2 fingers as you say, "5, 6").
- Say, **That's right. 4 plus 2** (show 2 fingers) **equals 6. The 6 is written here.** Point to the space where the 6 should be written.
- Say, **Let's do a subtraction problem. This says 4 minus 1. 4 minus 1 equals how much?**
- Say, **We can use our fingers to take away 1. 4** (show 4 fingers) **take away 1** (cover up 1 finger).
- Say, **4 minus 1 equals 3.**
- Say, **Now you can finish your paper. Remember to use your fingers if you do not know the answer.**
- Say, **When you finish, turn your paper over and fill in the missing numbers.**
- Hand out the Lesson 14 Activity Sheets and be prepared to help children with this new format. If a child finishes early, then check his or her work. If it is incorrect, then circle the error and have him or her correct it. If it is correct, then allow him or her to draw with the crayons.

Lesson 15

Learning Goals	Materials
Establish behavior boundaries	**COPY**
Count orally to 90, by tens to 100	Hundreds Chart
Build 15 as 10 and 5 ones	Cardinality Chart
Count and sequence to 15	Decade Card (10)
Number recognition 0–15	Unit Cards 1–9
Identify bigger/smaller number	Bigger/Smaller Cards
Find $n + 1$, $n - 1$	Teacher Number List
Make numbers 1–10 on fingers	Student Number List
Recognize quantities 0–10 on Ten Frames	Ten Frame Flash Cards
Write "before" and "after" numbers	Lesson 15 Activity Sheet
Solve equations: Vertical, +1, –1	**GATHER**
	White board magnetic easel
	Sticky notes
	15 interlocking blocks
	Pencils without erasers and crayons
	PREPARE
	Put student names on Lesson 15 Activity Sheets.
	Number Recognition Cards (0–15); *See Chapter 1 for instructions on making Number Recognition Cards.*

ESTABLISHING BEHAVIOR BOUNDARIES

- Say, **Show me you are ready to learn.** Wait for proper posture. **Exactly! That shows me you are ready!**

COUNTING WARM-UP

MATERIALS: Hundreds Chart, white board magnetic easel

- Put up the Hundreds Chart on the white board magnetic easel.
- Say, **Last time we counted 80. Today we are going to count even higher!**
- Gesture to the Hundreds Chart and say, **See how the numbers are lined up. Each row has 10 blocks. We can count by tens down this side.** Point to the far right column.
- Say, **Let's count by tens together. 10, 20, 30, 40, 50. Let's keep going. 60, 70, 80, 90, 100.** Draw your finger across each row and end on the decade as you count.
- Say, **We can count higher now that we know the pattern! Let's start at 65. I will start. 65.** Point to the child to your right. Use the Hundreds Chart for reference and error correction. Take a turn yourself.
- Continue to count to 90.

Hundreds Chart

Number Sense Interventions Lesson 15   93

MAGIC NUMBER ACTIVITIES (Magic Number is 15)

★ Making the Number 15

MATERIALS: 10 Decade Card, 5 Unit Card, 15 interlocking blocks

- Put down 15 unattached interlocking blocks. Say, **Here are some blocks. Let's count them together. You count on your fingers while I touch the blocks. Remember to use your inside voices. No shouting. Put your hands up so I can see them.**
- Count to 10, making a stick of 10 as you count. The children will have run out of fingers at 10. Hold up your fingers and say, **10 fingers! 10 blocks** (hold up the stick of 10 blocks).
- Say, **Let's keep counting.** Hold up the stick of 10 and as you say, "10." Touch each single block as you say, "11, 12, 13, 14, 15."
- Say, **Let's show that on our fingers. 10** (hold up 10 fingers then put down hands and begin counting on fingers again), **11, 12, 13, 14, 15.** Make sure children are counting correctly.
- Circle all the blocks and say, **How many blocks do we have altogether?** Wait for a response. **That is right. There are 15 blocks altogether.**
- Hold up the 10 Decade Card and say, **What number is this?** Wait for a response. Say, **Yes, the number is 10.**

10
Decade Card

- Say, **I will put the number 10 right next to the 10 blocks.** Lay the stick down and put down the 10 Decade Card next to the stick.
- Hold up the 5 Unit Card and say, **What number is this?** Wait for a response. **That is right. This is the number 5. I will put the number 5 right next to the 5 blocks.**

5
Unit Card

- The table should look like the following.

- Say, **Now watch what I do.** Put the 5 Unit Card over the 0 on the Decade Card while saying, **10 and 5 more is 15. This is the number 15. What number is this?** Wait for a response.
- Say, **That is right. This is the number 15. See the 10 hiding under the 5?** Lift up the 5 Unit Card. **15 is 10** (show the 10) **and 5 more** (add the 5 Unit Card on top).
- Then say, **10** (run your finger up the stick of 10) **and 5 more is 11, 12, 13, 14, 15** (touch each single block as you say, "11, 12, 13, 14, 15"). Repeat.
- Then say, **10** (run your finger up the stick of 10) **and 5 more** (point to the 5 blocks) **is 15.** Repeat.

★ Sequencing and Number Recognition 0-15

MATERIALS: Number Recognition Cards 0-15

- Say, **Say the numbers with me as I put them down. Do not go ahead of me. Ready? 1, 2, 3 ... 15!** Put the cards in 2 rows as if building a Hundreds Chart.
- Hold up the Number Recognition Card 15 and say, **What number is this?** Wait for a response.
- Say, **That is right. This is the number 15. Our Magic Number today is 15.**
- Say, **We are going to put the 15 here, under the 5. 15 is 10** (run your finger along the number list) **and 5 more** (point to the Number Recognition Card 15). Repeat.
- Say, **Let's play our game!**
- Go around the group, showing each child a different number from 0 to 15 (not in order). Make sure all children can see the number. Go around the group three times.

Number Recognition Cards

BIGGER/SMALLER

MATERIALS: Cardinality Chart, Bigger/Smaller Cards, white board magnetic easel

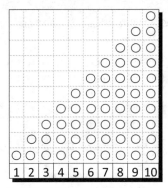

Cardinality Chart [number chart]

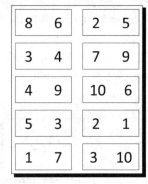

Bigger/Smaller Cards

- Show the Cardinality Chart only for error correction.
- Say, **I will show you a card with 2 numbers on it, and you tell me which number is bigger.**
- Go around the group twice, showing the Bigger/Smaller Cards in random order, saying, **Which number is bigger or more?**
- Say, **Now we are going to find smaller numbers. I will show you a card with 2 numbers on it, and you tell me which number is smaller.**
- Go around the group twice, showing the Bigger/Smaller Cards in random order, saying, **Which number is smaller or less?** Use the same error correction.

NUMBER LIST—NUMBER +1, −1

MATERIALS: Teacher Number List, Student Number List, Unit Cards (1-9), white board magnetic easel

- Put down the Teacher Number List.
- Say, **Now we are going to play our plus one game! You pick a card from this pile and we will add one to it. I will let you each have a number list to look at.** Pass out the Student Number Lists.

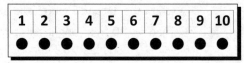

Teacher Number List

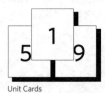

Unit Cards

- Let the children continue to take turns drawing Unit Cards from the pile. Say, **How much is ___ +1?**
- If the children hesitate or incorrectly respond, then point to the number list and show +1. For example, if the number was 5, then point to 5 and say, "5 plus 1" while moving up to 6.
- Say, **Now we are going to play a different game! You pick a card from this pile and we will minus one from it. Use your number list to find the number one down.**
- Let the children continue to take turns drawing Unit Cards from the pile. Say, **How much is ___ −1?**
- If the children hesitate or incorrectly respond, then point to the number list and show −1. For example, if the number was 5, then point to 5 and say, "5 minus 1" while moving down to 4.

SUBITIZING QUANTITIES (1-10) ACTIVITIES

★ Finger Automaticity

MATERIALS: Number Recognition Cards 1-10

- Say, **Let's play our finger game. I will show you a card, and you show me the number on your fingers. Be as quick as you can!**
- Shuffle the Number Recognition Cards and show them one at a time. Go around the group three times. Correct errors.

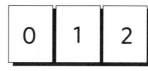

Number Recognition Cards

★ Recognizing Sets

MATERIALS: Ten Frame Flash Cards

- Say, **Now I am going to hold up a card with some circles on it, and I want you to tell me how many circles are on the card. Try to tell me as fast as you can without counting. I will point to you for your turn, okay? Everyone else, say it in your mind.**
- Go through the Ten Frame Flash Cards in random order. Go around the group three times. Correct errors.

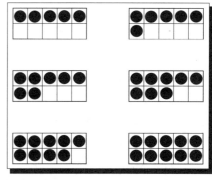

Ten Frame Flash Cards

WRITTEN NUMBERS

MATERIALS: White board magnetic easel, Lesson 15 Activity Sheets, pencils, crayons

- Put a copy of the Lesson 15 Activity Sheet on the white board magnetic easel.
- Say, **Sometimes number sentences are written this way, up and down. This says 3 plus 1. 3 plus 1 equals how much?** Wait for a response.
- Say, **All the problems are plus 1 or minus 1. Remember to go up 1 or down 1 on the number list. I will keep my list here so you can look at it.**
- Say, **When you finish, turn your paper over and fill in the missing numbers.**
- Hand out the Lesson 15 Activity Sheets and be prepared to help children with this new format. If a child finishes early, then check his or her work. If it is incorrect, then circle the error and have him or her correct it. If it is correct, then allow him or her to draw with the crayons.

Lesson 15 Activity Sheet

Lesson 16

Learning Goals

Establish behavior boundaries
Count orally to 100, by tens to 100
Build 16 as 10 and 6 ones
Count and sequence to 16
Number recognition 0–16
Before and after (1–10)
Find $n + 1$, $n - 1$
Make numbers 1–10 on fingers
Draw story problems

Materials

COPY
Hundreds Chart
Decade Card (10)
Unit Cards 1–9
Teacher Number List
Student Number List
Ten Frame Flash Cards
Lesson 16 Activity Sheet

GATHER
White board magnetic easel
16 interlocking blocks
Dry erase marker and eraser
Pencils without erasers, crayons

PREPARE
Put student names on Lesson 16 Activity Sheets.
Number Recognition Cards (0–16); *See Chapter 1 for instructions on making Number Recognition Cards.*

ESTABLISHING BEHAVIOR BOUNDARIES

- Say, **Show me you are ready to learn.** Wait for proper posture. **Exactly! That shows me you are ready!**

COUNTING WARM-UP

MATERIALS: Hundreds Chart, white board magnetic easel

- Put up the Hundreds Chart on the white board magnetic easel.
- Say, **Last time we counted to 90. Today we are going to count even higher!**
- Gesture to the Hundreds Chart and say, **See how the numbers are lined up. Each row has 10 blocks. We can count by tens down this side.** Point to the far right column.
- Say, **Let's count by tens together. 10, 20, 30, 40, 50. Let's keep going. 60, 70, 80, 90, 100.** Draw your finger across each row and end on the decade as you count.
- Say, **We can count higher now that we know the pattern! Let's start at 76. I will start. 76.** Point to the child to your right. Use the Hundreds Chart for reference and error correction. Take a turn yourself.
- Continue to 100.

1	2	3	4	5	6	7	8	9	10
11	12	13	14	15	16	17	18	19	20
21	22	23	24	25	26	27	28	29	30
31	32	33	34	35	36	37	38	39	40
41	42	43	44	45	46	47	48	49	50
51	52	53	54	55	56	57	58	59	60
61	62	63	64	65	66	67	68	69	70
71	72	73	74	75	76	77	78	79	80
81	82	83	84	85	86	87	88	89	90
91	92	93	94	95	96	97	98	99	100

Hundreds Chart

Number Sense Interventions Lesson 16   97

MAGIC NUMBER ACTIVITIES (Magic Number is 16)

★ Making the Number 16

MATERIALS: 10 Decade Card, 6 Unit Card, 16 interlocking blocks

- Put down 16 unattached interlocking blocks. Say, **Here are some blocks. Let's count them together. You count on your fingers while I touch the blocks. Remember to use your inside voices. No shouting. Put your hands up so I can see them.**
- Count to 10, making a stick of 10 as you count. The children will have run out of fingers at 10. Hold up your fingers and say, **10 fingers! 10 blocks** (hold up the stick of 10 blocks).
- Say, **Let's keep counting.** Hold up the stick of 10 and say, "10." Touch each single block as you say, "11, 12, 13, 14, 15, 16."
- Say, **Let's show that on our fingers. 10** (hold up 10 fingers then put down hands and begin counting on fingers again), **11, 12, 13, 14, 15, 16.** Make sure children are counting correctly.
- Circle all the blocks and say, **How many blocks do we have altogether?** Wait for a response. **That is right. There are 16 blocks altogether.**
- Hold up the 10 Decade Card and say, **What number is this?** Wait for a response. Say, **Yes, the number is 10.**

Decade Card

- Say, **I will put the number 10 right next to the 10 blocks.** Lay the stick down and put down the 10 Decade Card next to the stick. Set up table as in previous lessons.
- Hold up the 6 Unit Card and say, **What number is this?** Wait for a response. **That is right. This is the number 6. I will put the number 6 right next to the 6 blocks.**
- Say, **Now watch what I do.** Put the Unit Card 6 over the 0 on the Decade Card while saying, **10 and 6 more is 16. This is the number 16. What number is this?** Wait for a response.

Unit Card

- Say, **That is right. This is the number 16. See the 10 hiding under the 6** (lift up the 6 Unit Card)? **16 is 10 and 6 more** (show the 10 then add the 6 on top as you are saying, "and 6 more").
- Then say, **10** (run your finger up the stick of 10) **and 6 more is 11, 12, 13, 14, 15, 16.** Touch each single block as you say, "11, 12, 13, 14, 15, 16". Repeat.
- Then say, **10** (run your finger up the stick of 10) **and 6 more** (point to the 6 blocks) **is 16.** Repeat.

★ Sequencing and Number Recognition 0-16

MATERIALS: Number Recognition Cards 0-16

Number Recognition Cards

- Say, **Say the numbers with me as I put them down. Do not go ahead of me. Ready? 1, 2, 3 ... 16!** Put the cards in two rows as if building a Hundreds Chart.
- Hold up the Number Recognition Card 16 and say, **What number is this?** Wait for a response.
- Say, **That is right. This is the number 16. Our Magic Number today is 16.**
- Say, **We are going to put the 16 here, under the 6. 16 is 10** (run your finger along the number list) **and 6 more** (point to the Number Recognition Card 16). Repeat.
- Say, **Let's play our game!**
- Go around the group, showing each child a different number from 0 to 16 (not in order). Make sure all children can see the number. Go around the group three times.

★ Before and After

MATERIALS: Number Recognition Cards 1-10

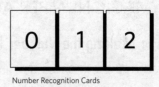
Number Recognition Cards

- Say, **I am going to show you a number, and I want you to tell me what number comes right after it. We will take turns, so do not call out if it is not your turn. Think carefully and tell me what number comes after.**
- Show the Number Recognition Cards 1-10 in random order. As you hold up a card say, **What number comes right after _____?**
- Give each child two turns. Repeat with before numbers.
- Error correction: If the child gives a wrong answer, then put the card on the table and say, **What number comes after/before____** and point to the space after as in previous activities.

NUMBER LIST—NUMBER +1, −1

MATERIALS: Teacher Number List, Student Number List, Unit Cards 1-9

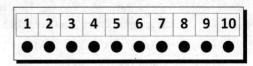
Teacher Number List

- Put down the Teacher Number List.
- Say, **Now we are going to play our plus one game! You pick a card from this pile and we will add one to it. I will let you each have a number list to look at.** Pass out Student Number Lists.
- Let the children continue to take turns drawing Unit Cards from the pile. Say, **How much is ___ + 1?**
- If the children hesitate or incorrectly respond, then point to the number list and show + 1. For example, if the number was 5, then point to 5 and say, "5 plus 1" while moving up to 6.
- Say, **Now we are going to play a different game! You pick a card from this pile and we will minus one from it. Use your number list to find the number one down.**
- Let the children continue to take turns drawing Unit Cards from the pile. Say, **How much is ___ − 1?**
- If the children hesitate or incorrectly respond, then point to the number list and show − 1. For example, if the number was 5, then point to 5 and say, "5 minus 1" while moving down to 4.

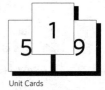
Unit Cards

SUBITIZING QUANTITIES (1-10) ACTIVITIES

★ Finger Automaticity

MATERIALS: Number Recognition Cards 1-10

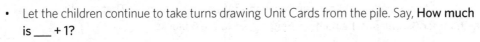
Number Recognition Cards

- Say, **Let's play our finger game. I will show you a card, and you show me the number on your fingers. Be as quick as you can!**
- Shuffle the Number Recognition Cards and show them one at a time. Go around the group three times. Correct errors.

Number Sense Interventions Lesson 16   99

★ Recognizing Sets

MATERIALS: Ten Frame Flash Cards

- Say, **Now I am going to hold up a card with some circles on it, and I want you to tell me how many circles are on the card. Try to tell me as fast as you can without counting. I will point to you for your turn, okay? Everyone else, say it in your mind.**
- Go through the Ten Frame Flash Cards in random order. Go around the group three times. Correct errors.

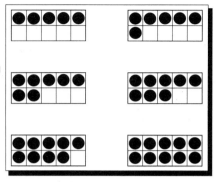

Ten Frame Flash Cards

DRAWING STORY PROBLEMS

MATERIALS: Lesson 16 Activity Sheets, pencils, erasers, white board magnetic easel, dry erase marker and eraser

- Say, **Now I am going to tell some stories, and I want you to use a drawing to help you solve the problem. The problem may be about animals or cookies, but you should not draw animals or cookies. You can keep track by drawing sticks or circles.**
- Say, **Let's try some together.**

2 + 2 = 4

- Hand out the Lesson 16 Activity Sheets and say, **Put your finger in the box with the smiley face. Listen to this story. Sue has 2 pennies. Let's draw 2 circles to stand for 2 pennies.** Model on the white board magnetic easel.
- Say, **Her brother gives her 2 more pennies. Draw 2 more circles.** Model on the white board magnetic easel.
- Say, **How many pennies does Sue have now? Can you see the answer? 2 plus 2 is how many?** Wait for a response. Gesture to your drawing. Circle each set of 2 and then the whole picture. Encourage subitizing, not counting.
- Say, **Write 4 in the box for 4 pennies.** Check students' drawings. Discourage complicated drawings, and encourage models such as tallies or circles.

Lesson 16 Activity Sheet

5 + 1 = 6

- Say, **Put your finger in the box with the heart. Listen to this story. There were 5 pigs in the pen. Draw sticks to stand for 5 pigs.** Model on the white board magnetic easel.
- Say, **The farmer brought 1 more pig into the pen. Draw a stick for the 1 more pig.** Model on the white board magnetic easel.
- Say, **How many pigs are in the pen now? 5 and 1 more is how much?** Wait for a response. **Write 6 in the box for 6 pigs.**
- Say, **Now I am going to tell some stories with take away or minus, and I want you to use a drawing to help you solve the problem. Let's try some together.**

$5 - 2 = 3$

- Say, **Put your finger in the box with the sun. Listen to this story. Sue has 5 pennies. Draw circles to stand for 5 pennies.** Model on the white board magnetic easel.
- Say, **She gave her brother 2 of her pennies. Cross out the 2 pennies she gave her brother.** Model on white board magnetic easel.
- Say, **How many pennies does Sue have now? Tell me without counting.** Wait for a response. Check students' drawings.
- Say, **That is right. 3 pennies! Write your answer in the box.**

$6 - 4 = 2$

- Say, **Put your finger in the box with the triangle. Listen to this story. There were 6 pigs in the mud. Draw sticks to stand for 6 pigs.** Model on the white board magnetic easel.
- Say, **Then, 4 of them ran into the barn! Cross out the 4 pigs that ran into the barn.** Model on the white board magnetic easel.
- Say, **How many pigs are in the mud now? Tell me without counting?** Wait for a response. Check students' drawings.
- Say, **That is right. 2 pigs. Write your answer in the box.**

Lesson 17

Learning Goals	Materials
Establish behavior boundaries	**COPY**
Count orally to 100, by tens to 100	Cardinality Chart
Build 17 as 10 and 7 ones	Hundreds Chart
Count and sequence to 17	Decade Card (10)
Number recognition 0–17	Unit Card (7)
Identify bigger/smaller number	Bigger/Smaller Cards
Make numbers 1–10 on fingers	Ten Frame Flash Cards
Draw story problems	Lesson 17 Activity Sheet
	GATHER
	White board magnetic easel
	17 interlocking blocks
	Dry erase marker and eraser
	Pencils without erasers and crayons
	PREPARE
	Put student names on Lesson 17 Activity Sheets.
	Number Recognition Cards (0–17); See Chapter 1 for instructions on making Number Recognition Cards.

Number Sense Interventions Lesson 17   101

ESTABLISHING BEHAVIOR BOUNDARIES

- Say, **Show me you are ready to learn.** Wait for proper posture. **Exactly! That shows me you are ready!**

COUNTING WARM-UP

MATERIALS: Hundreds Chart, white board magnetic easel

- Put up the Hundreds Chart, and use it for reference when needed.
- Say, **Today we are going to count higher numbers. I am going to start counting and then you keep going.**
- Start with 25, 26, and let children continue until they all have 2 turns.
- Repeat with 47, 48; 54, 55; 66, 67; 78, 79. Use error correction from previous lessons.

1	2	3	4	5	6	7	8	9	10
11	12	13	14	15	16	17	18	19	20
21	22	23	24	25	26	27	28	29	30
31	32	33	34	35	36	37	38	39	40
41	42	43	44	45	46	47	48	49	50
51	52	53	54	55	56	57	58	59	60
61	62	63	64	65	66	67	68	69	70
71	72	73	74	75	76	77	78	79	80
81	82	83	84	85	86	87	88	89	90
91	92	93	94	95	96	97	98	99	100

Hundreds Chart

MAGIC NUMBER ACTIVITIES (Magic Number is 17)

★ Making the Number 17

MATERIALS: 10 Decade Card, 7 Unit Card, 17 interlocking blocks

- Put down 17 unattached interlocking blocks. Say, **Here are some blocks. Let's count them together. You count on your fingers while I touch the blocks. Remember to use your inside voices. No shouting. Put your hands up so I can see them.**
- Count to 10, making a stick of 10 as you count. The children will have run out of fingers at 10. Hold up your fingers and say, **10 fingers! 10 blocks** (hold up the stick of 10 blocks).
- Say, **Let's keep counting.** Hold up the stick of 10 and say, "10." Touch each single block as you say, "11, 12, 13, 14, 15, 16, 17."
- Say, **Let's show that on our fingers. 10** (hold up 10 fingers then put down hands and begin counting on fingers again), **11, 12, 13, 14, 15, 16, 17.** Make sure children are counting correctly.
- Circle all the blocks and say, **How many blocks do we have altogether?** Wait for a response. **That is right. There are 17 blocks altogether.**
- Hold up the 10 Decade Card and say, **What number is this?** Wait for a response. Say, **Yes, the number is 10.**

Decade Card

- Say, **I will put the number 10 right next to the 10 blocks.** Lay the stick down and put down the 10 Decade Card next to the stick. Set up the table as in previous lessons.
- Hold up the 7 Unit Card and say, **What number is this?** Wait for a response. **That is right. This is the number 7. I will put the number 7 right next to the 7 blocks.**
- Say, **Now watch what I do.** Put the 7 Unit Card over the 0 on the Decade Card while saying, **10 and 7 more is 17. This is the number 17. What number is this?** Wait for a response.

Unit Card

- Say, **That is right. This is the number 17. See the 10 hiding under the 7** (lift up the 7 Unit Card)? **17 is 10 and 7 more** (show the 10 then add the 7 on top as you are saying, "and 7 more").
- Then say, **10** (run your finger up the stick of 10) **and 7 more is 11, 12, 13, 14, 15, 16, 17** (touch each single block as you say, "11, 12, 13, 14, 15, 16, 17"). Repeat.
- Then say, **10** (run your finger up the stick of 10) **and 7 more** (point to the 7 blocks) **is 17.** Repeat.

★ Sequencing and Number Recognition 0-17

MATERIALS: Number Recognition Cards 0-17

- Say, **Say the numbers with me as I put them down. Do not go ahead of me. Ready? 1, 2, 3 … 17!** Put the cards in 2 rows as if building a Hundreds Chart.
- Hold up the Number Recognition Card 17 and say, **What number is this?** Wait for a response.
- Say, **That is right. This is the number 17. Our Magic Number today is 17.**
- Say, **We are going to put the 17 here, under the 7. 17 is 10** (run your finger along the number list) **and 7 more** (point to the Number Recognition Card 17). Repeat.
- Say, **Let's play our game!**
- Go around the group, showing each child a different number from 0 to 17 (not in order). Make sure all children can see the number. Go around the group three times.

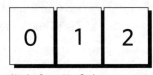

Number Recognition Cards

BIGGER/SMALLER

MATERIALS: Cardinality Chart, Bigger/Smaller Cards, white board magnetic easel

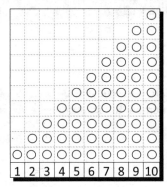

Cardinality Chart [number chart]

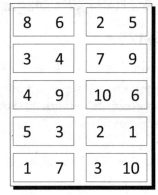

Bigger/Smaller Cards

- Show the Cardinality Chart only for error correction.
- Say, **I will show you a card with 2 numbers on it, and you tell me which number is bigger.**
- Go around the group twice, showing the Bigger/Smaller Cards in random order, saying, **Which number is bigger or more?**
- Say, **Now we are going to find smaller numbers. I will show you a card with 2 numbers on it, and you tell me which number is smaller.**
- Go around the group twice, showing the Bigger/Smaller cards in random order, saying, **Which number is smaller or less?**

SUBITIZING QUANTITIES (1-10) ACTIVITIES

★ Finger Automaticity

MATERIALS: Number Recognition Cards 1-10

- Say, **Let's play our finger game. I will show you a card, and you show me the number on your fingers. Be as quick as you can!**
- Shuffle the Number Recognition Cards and show them one at a time. Go around the group three times. Correct errors.

Number Recognition Cards

Number Sense Interventions Lesson 17   103

★ Recognizing Sets

MATERIALS: Ten Frame Flash Cards

- Say, **Now I am going to hold up a card with some circles on it, and I want you to tell me how many circles are on the card. Try to tell me as fast as you can without counting. I will point to you for your turn, okay? Everyone else, say it in your mind.**

- Go through the Ten Frame Flash Cards in random order. Go around the group three times. Correct errors.

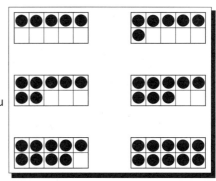
Ten Frame Flash Cards

DRAWING STORY PROBLEMS

MATERIALS: Lesson 17 Activity Sheets, pencil, erasers, white board magnetic easel, dry erase marker and eraser

- Say, **I am going to tell some stories, and I want you to use a drawing to help you solve the problem. The problem may be about animals or cookies, but you should not draw animals or cookies. You can keep track by drawing sticks or circles.** Do not model on the white board magnetic easel until they have completed their drawings.

2 + 3 = 5

- Hand out the Lesson 17 Activity Sheets and say, **Put your finger in the box with the arrow. Listen to this story. Bill has 2 apples and John has 3 apples. How many apples do they have altogether? Make your drawing.**

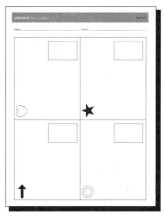
Lesson 17 Activity Sheet

- Say, **How much is 2 plus 3? Write your answer in the box.** You may repeat the problem but do not say just the numbers. Say the problem.

- Say, **Now I will make a drawing and you can check yours.** Repeat the problem as you draw.

- Say, **What number did you write in the box? How many apples do they have altogether?** Wait for a response.

3 + 1 = 4

- Say, **Put your finger in the box with the heart. Listen to this story. There were 3 pigs in the pen. The farmer brought 1 more pig into the pen. How many pigs are in the pen now? Try to answer without counting. Write your answer in the box.**

- Model on the white board magnetic easel.

- Say, **Now I am going to tell some stories with take away or minus. Remember to cross out what is taken away! Ready?**

6 − 3 = 3

- Say, **Put your finger in the box with the sun. Listen to this story. Sue has 6 pennies. She gave her brother 3 of her pennies. Cross out the 3 pennies she gave her brother. How many pennies does Sue have now? Try to answer without counting. Write your answer in the box.**

- Model on the white board magnetic easel.

$5 - 3 = 2$

- Say, **Put your finger in the box with the star. Listen to this story. The balloon man had 5 balloons. Then 3 of them popped! Cross out the 3 balloons that popped. How many balloons does the balloon man have now? Try to answer without counting. Write your answer in the box.**
- Model on the white board magnetic easel.

Lesson 18

Learning Goals	Materials
Establish behavior boundaries Count orally to 100, by tens to 100 Build 18 as 10 and 8 ones Count and sequence to 18 Number recognition 0–18 Before and after (1–10) Find $n + 1$, $n - 1$ Make numbers 1–10 on fingers Draw story problems	**COPY** Hundreds Chart Decade Card (10) Unit Cards 1–9 Ten Frame Flash Cards Teacher Number List Student Number List Activity 18 Hundreds Chart Lesson 18 Activity Sheet **GATHER** White board magnetic easel 18 interlocking blocks Marker Pencils without erasers and crayons **PREPARE** Put student names on Lesson 18 Activity Sheets. Number Recognition Cards (0–18); *See Chapter 1 for instructions on making Number Recognition Cards.*

ESTABLISHING BEHAVIOR BOUNDARIES

- Say, **Show me you are ready to learn.** Wait for proper posture. **Exactly! That shows me you are ready!**

COUNTING WARM-UP

MATERIALS: Hundreds Chart, white board magnetic easel

- Put up the Hundreds Chart to use for reference when needed.
- Say, **Today we are going to count higher numbers. I am going to start counting and then you keep going.**
- Start with 57, 58, and let the children continue until they all have 2 turns.
- Repeat with 64, 65; 76, 77; 88, 89. Use error correction from previous lessons.

Hundreds Chart

Number Sense Interventions Lesson 18   105

HUNDREDS CHART ACTIVITY

MATERIALS: Activity 18 Hundreds Chart, marker

- Gesture to the Activity 18 Hundreds Chart and say, **Who can find a pattern on this number chart?**
- Continue to give children opportunities to find patterns. Guide them to see both row and column patterns (e.g., rows have the same decade number; rows go in order: 1, 2, 3; columns have the same unit number).
- Display the Activity 18 Hundreds Chart. Say, **Here is a Hundreds Chart, but some of the numbers are missing. Let's figure out the missing numbers.**
- Point to the first empty space and say, **What number is missing here?** If students have a hard time getting started, then count up the row until they get to the number.
- Check it by looking at the column to be sure every number ends in the same number.
- Fill in numbers one row at a time.

1		3	4	5	6	7	8	9	10
11	12	13	14		16	17	18	19	20
21	22		24	25	26	27	28	29	30
31	32	33	34	35	36	37		39	40
41	42	43	44	45		47	48	49	50
51	52	53		55	56	57	58	59	60
61	62	63	64	65	66		68	69	70
71	72	73	74	75	76	77		79	80
81	82	83	84	85	86	87	88		90
91	92	93	94	95	96	97	98	99	

Activity 18 Hundreds Chart

MAGIC NUMBER ACTIVITIES (Magic Number is 18)

★ Making the Number 18

MATERIALS: 10 Decade Card, 8 Unit Card, 18 interlocking blocks

- Put down 18 unattached interlocking blocks. Say, **Here are some blocks. Let's count them together. You count on your fingers while I touch the blocks. Remember to use your inside voices. No shouting. Put your hands up so I can see them.**
- Count to 10, making a stick of 10 as you count. The children will have run out of fingers at 10. Hold up your fingers and say, **10 fingers! 10 blocks** (hold up the stick of 10 blocks).
- Say, **Let's keep counting.** Hold up the stick of 10 and say, "10." Touch each single block as you say, "11, 12, 13, 14, 15, 16, 17, 18."
- Say, **Let's show that on our fingers. 10** (hold up 10 fingers then put down hands and begin counting on fingers again), **11, 12, 13, 14, 15, 16, 17, 18.** Make sure children are counting correctly.
- Circle all the blocks and say, **How many blocks do we have altogether?** Wait for a response. **That is right. There are 18 blocks altogether.**
- Hold up the 10 Decade Card and say, **What number is this?** Wait for a response. Say, **Yes, the number is 10.**
- Say, **I will put the number 10 right next to the 10 blocks.** Lay the stick down and put down the 10 Decade Card next to the stick. Set up the table as in previous lessons.
- Hold up the 8 Unit Card and say, **What number is this?** Wait for a response. **That is right. This is the number 8. I will put the number 8 right next to the 8 blocks.**
- Say, **Now watch what I do.** Put the Unit Card 8 over the 0 on the Decade Card while saying, **10 and 8 more is 18. This is the number 18. What number is this?** Wait for a response.
- Say, **That is right. This is the number 18. See the 10 hiding under the 8** (lift up the Unit Card 8)? **18 is 10 and 8 more** (Show the 10 then add the 8 on top as you are saying, "and 8 more").
- Then say, **10** (run your finger up the stick of 10) **and 8 more is 11, 12, 13, 14, 15, 16, 17, 18** (touch each single block as you say, "11, 12, 13, 14, 15, 16, 17, 18"). Repeat.
- Then say, **10** (run your finger up the stick of 10) **and 8 more** (point to the 8 blocks) **is 18.** Repeat.

10
Decade Card

8
Unit Card

★ Sequencing and Number Recognition 0-18

MATERIALS: Number Recognition Cards 0-18

Number Recognition Cards

- Say, **Say the numbers with me as I put them down. Do not go ahead of me. Ready? 1, 2, 3 ... 18.** Put the cards in 2 rows as if building a Hundreds Chart.
- Hold up the Number Recognition Card 18 and say, **What number is this?** Wait for a response.
- Say, **That is right. This is the number 18. Our Magic Number today is 18.**
- Say, **We are going to put the 18 here, under the 8. 18 is 10** (run your finger along the number list) **and 8 more** (point to the Number Recognition Card 18). Repeat.
- Say, **Let's play our game!**
- Go around the group, showing each child a different number from 0 to 18 (not in order). Make sure all children can see the number. Go around the group four times.

★ Before and After

MATERIALS: Number Recognition Cards 1-10

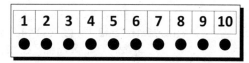

- Say, **I am going to show you a number, and I want you to tell me what number comes right after it. We will take turns, so do not call out if it is not your turn. Think carefully and tell me what number comes after.**
- Show Number Recognition Cards 1-10 in random order. As you hold up a card say, **What number comes right after _____?**
- Give each child two turns. Repeat with before numbers.
- Error correction: If the child gives a wrong answer, then put the card on the table and say, **What number comes after/before_____** and point to the space after as in previous activities.

NUMBER LIST—NUMBER +1, −1

MATERIALS: Teacher Number List, Student Number List, Unit Cards 1-9

- Put down the Teacher Number List.
- Say, **Now we are going to play our plus one game! You pick a card from this pile and we will add one to it. I will let you each have a number list to look at.** Pass out Student Number Lists.
- Let the children continue to take turns drawing cards from the pile. Say, **How much is ___ + 1?**
- If the children hesitate or incorrectly respond, then point to the number list and show + 1. For example, if the number was 5, then point to 5 and say, "5 plus 1" while moving up to 6.
- Say, **Now we are going to play a different game! You pick a card from this pile and we will minus one from it. Use your number list to find the number one down.**
- Let the children continue to take turns drawing Unit Cards from the pile. Say, **How much is ___ − 1?**
- If the children hesitate or incorrectly respond, then point to the number list and show − 1. For example, if the number was 5, then point to 5 and say, "5 − 1" while moving down to 4.

Number Sense Interventions Lesson 18   107

SUBITIZING QUANTITIES (1–10) ACTIVITIES

★ Finger Automaticity

MATERIALS: Number Recognition Cards 1–10

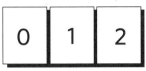
Number Recognition Cards

- Say, **Let's play our finger game. I will show you a card, and you show me the number on your fingers. Be as quick as you can!**
- Shuffle the Number Recognition Cards and show them one at a time. Go around three times. Correct errors.

★ Recognizing Sets

MATERIALS: Ten Frame Flash Cards

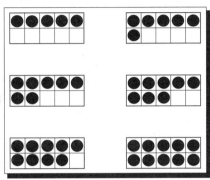
Ten Frame Flash Cards

- Say, **Now I am going to hold up a card with some circles on it, and I want you to tell me how many circles are on the card. Try to tell me as fast as you can without counting. I will point to you for your turn, okay? Everyone else, say it in your mind.**
- Go through the Ten Frame Flash Cards in random order. Go around the group three times. Correct errors.

DRAWING STORY PROBLEMS

MATERIALS: Lesson 18 Activity Sheets, pencils, erasers

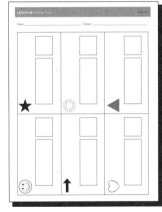
Lesson 18 Activity Sheet

Note: If there is not enough time to do this entire activity, do at least one addition and one subtraction. At the next lesson time, begin with a counting warm-up and some flash card activities, followed by the remaining story problems. Do not go ahead to Lesson 19 until all these story problems have been completed.

- Say, **I am going to tell some stories, and I want you to use a drawing to help you solve the problem. You can keep track by drawing sticks or circles, whichever is easy for you to draw. Listen carefully and do your best! Do not call out the answers. Just write them down on your paper.** Do not demonstrate. This can be used as an assessment.

2 + 4 = 6

- Hand out the Lesson 18 Activity Sheets and say, **Put your finger in the box with the smiley face. Listen to this story. Bill has 2 apples. John has 4 apples. How many apples do they have altogether? Do not call out the answer. Just write it in the box.**

5 + 2 = 7

- Say, **Put your finger in the box with the arrow. Listen to this story. There were 5 pigs in the pen. The farmer brought 2 more pigs into the pen. How many pigs are in the pen now? Do not call out the answer. Just write it in the box.**

3 + 4 = 7

- Say, Put your finger in the box with the heart. Listen to this story. Bill has 3 apples. John has 4 apples. How many apples do they have altogether? Do not call out the answer. Just write it in the box.
- Say, Now I am going to tell some stories with take away or minus. Remember to cross out what is taken away!

6 − 2 = 4

- Say, Put your finger in the box with the star. Listen to this story. Sue has 6 pennies. She gave her brother 2 of her pennies. Cross out the 2 pennies she gave her brother. How many pennies does Sue have now? Do not call out the answer. Just write it in the box.

5 − 3 = 2

- Say, Put your finger in the box with the sun. Listen to this story. 5 frogs were sitting on a log. 3 jumped off into the water! How many frogs are on the log now? Do not call out the answer. Just write it in the box.

7 − 7 = 0

- Say, Put your finger in the box with the triangle. Listen to this story. There were 7 pigs in the pen. 7 of them ran out of the pen and into the barn! How many pigs are in the pen now? Do not call out the answer. Just write it in the box.

Lesson 19

Learning Goals

Establish behavior boundaries
Count orally to 100, by tens to 100
Build 19 as 10 and 9 ones
Count and sequence to 19
Number recognition 0–19
Identify missing numbers on Hundreds Chart
Identify bigger/smaller number
Make numbers 1–10 on fingers
Ten Frames: Nickel and pennies
Number operations on fingers: Counting-on
Use number sentences to solve story problems

Materials

COPY
Hundreds Chart
Activity 19 Hundreds Chart
Cardinality Chart
Decade Card (10)
Unit Cards (9)
Teacher Ten Frame Mat
Bigger/Smaller Cards
Lesson 19 Activity Sheet

GATHER
White board magnetic easel
Marker
19 interlocking blocks
Dry erase marker and eraser
8 red dots and 2 yellow dots
Pencils without erasers

PREPARE
Put student names on Lesson 19 Activity Sheets
Number Recognition Cards (0–19); *See Chapter 1 for instructions on making Number Recognition Cards.*
10 Penny Magnets; *See Chapter 1 for instructions on making Penny Magnets.*
Nickel Magnet Strip; *See Chapter 1 for instructions on making a Nickel Magnet Strip.*
Vertical Flash Cards: 6 + 2, 2 + 6, 7 + 3, 4 + 6, 2 + 7; *See Chapter 1 for more information on Vertical Flash Cards.*

ESTABLISHING BEHAVIOR BOUNDARIES

- Say, **Show me you are ready to learn.** Wait for proper posture. **Exactly! That shows me you are ready!**

COUNTING WARM-UP

MATERIALS: Hundreds Chart, white board magnetic easel

- Put up the Hundreds Chart to use for reference when needed.
- Say, **Today we are going to count higher numbers. I am going to start counting and then you keep going.**
- Start with 25, 26, and let children continue until they all have 2 turns.
- Repeat with 67, 68; 74, 75; 86, 87; 95, 96. Use error correction from previous lessons.

1	2	3	4	5	6	7	8	9	10
11	12	13	14	15	16	17	18	19	20
21	22	23	24	25	26	27	28	29	30
31	32	33	34	35	36	37	38	39	40
41	42	43	44	45	46	47	48	49	50
51	52	53	54	55	56	57	58	59	60
61	62	63	64	65	66	67	68	69	70
71	72	73	74	75	76	77	78	79	80
81	82	83	84	85	86	87	88	89	90
91	92	93	94	95	96	97	98	99	100

Hundreds Chart

HUNDREDS CHART ACTIVITY

MATERIALS: Activity 19 Hundreds Chart, marker

- Gesture to the Activity 19 Hundreds Chart and say, **Who can find a pattern on this number chart?**
- Continue to give children opportunities to find patterns. Guide them to see both row and column patterns (e.g., rows have the same decade number; rows go in order: 1, 2, 3; columns have the same unit number).
- Display the Activity 19 Hundreds Chart. Say, **Here is a Hundreds Chart, but some of the numbers are missing. Let's figure out the missing numbers.**
- Point to the first empty space and say, **What number is missing here?** If students have a hard time getting started, then count up the row until they get to the number.
- Check it by looking at the column to be sure every number ends in the same number.
- Fill in the numbers one row at a time.

Activity 19 Hundreds Chart

MAGIC NUMBER ACTIVITIES (Magic Number is 19)

★ Making the Number 19

MATERIALS: 10 Decade Card, 9 Unit Card, 19 interlocking blocks

- Put down 19 unattached interlocking blocks. Say, **Here are some blocks. Let's count them together. You count on your fingers while I touch the blocks. Remember to use your inside voices. No shouting. Put your hands up so I can see them.**
- Count to 10, making a stick of 10 as you count. The children will have run out of fingers at 10. Hold up your fingers and say, **10 fingers! 10 blocks** (hold up the stick of 10 blocks).
- Say, **Let's keep counting.** Hold up the stick of 10 and say, "10." Touch each single block as you say, "11, 12, 13, 14, 15, 16, 17, 18, 19."
- Say, **Let's show that on our fingers. 10** (hold up 10 fingers then put down hands and begin counting on fingers again), **11, 12, 13, 14, 15, 16, 17, 18, 19.** Make sure children are counting correctly.
- Circle all the blocks and say, **How many blocks do we have altogether?** Wait for a response. **That is right. There are 19 blocks altogether.**
- Hold up the 10 Decade Card and say, **What number is this?** Wait for a response. Say, **Yes, the number is 10.**
- Say, **I will put the number 10 right next to the 10 blocks.** Lay the stick down and put down the 10 Decade Card next to the stick. Set up table as in previous lessons.
- Hold up the 9 Unit Card and say, **What number is this?** Wait for a response. **That is right. This is the number 9. I will put the number 9 right next to the 9 blocks.**
- Say, **Now watch what I do.** Put the Unit Card 9 over the 0 on the Decade Card while saying, **10 and 9 more is 19. This is the number 19. What number is this?** Wait for a response.
- Say, **That is right. This is the number 19. See the 10 hiding under the 9** (lift up the Unit Card 9)? **19 is 10 and 9 more** (Show the 10 then add the 9 on top as you are saying, "and 9 more").
- Then say, **10** (run your finger up the stick of 10) **and 9 more is 11, 12, 13, 14, 15, 16, 17, 18, 19** (touch each single block as you say, "11, 12, 13, 14, 15, 16, 17, 18, 19"). Repeat.
- Then say, **10** (run your finger up the stick of 10) **and 9 more** (point to the 9 blocks) **is 19.** Repeat.

Decade Card

Unit Card

Number Sense Interventions Lesson 19   111

★ Sequencing and Number Recognition 0-19

MATERIALS: Number Recognition Cards 0-19

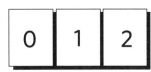

- Say, **Say the numbers with me as I put them down. Do not go ahead of me. Ready? 1, 2, 3 ... 19.** Put the cards in 2 rows as if building a Hundreds Chart.
- Hold up the Number Recognition Card 19 and say, **What number is this?** Wait for a response.
- Say, **That is right. This is the number 19. Our Magic Number today is 19.**
- Say, **We are going to put the 19 here, under the 9. 19 is 10** (run your finger along the number list) **and 9 more** (point to the Number Recognition Card 19). Repeat.
- Say, **Let's play our game!**
- Go around the group, showing each child a different number from 0 to 19 (not in order). Make sure all children can see the number. Go around the group four times.

BIGGER/SMALLER

MATERIALS: Cardinality Chart, Bigger/Smaller Cards, white board magnetic easel

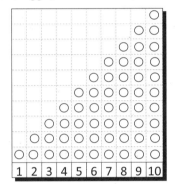
Cardinality Chart [number chart]

Bigger/Smaller Cards

- Show the Cardinality Chart only for error correction.
- Say, **I will show you a card with 2 numbers on it, and you tell me which number is bigger.**
- Go around the group twice, showing the Bigger/Smaller Cards in random order, saying, **Which number is bigger or more?**
- Say, **Now we are going to find smaller numbers. I will show you a card with 2 numbers on it, and you tell me which number is smaller.**
- Go around the group twice, showing the Bigger/Smaller Cards in random order, saying, **Which number is smaller or less?**

SUBITIZING QUANTITIES (1-10) ACTIVITIES

★ Finger Automaticity

MATERIALS: Number Recognition Cards 1-10

- Say, **Let's play our finger game. I will show you a card, and you show me the number on your fingers. Be as quick as you can!**
- Shuffle the Number Recognition Cards and show them one at a time. Go around the group three times. Correct errors.

★ Nickels and Pennies

MATERIALS: White board magnetic easel, Teacher Ten Frame Mat, Nickel Magnet Strip, 10 Penny Magnets

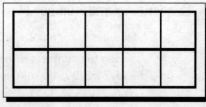
Ten Frame Mat

- Put the Teacher Ten Frame Mat on the white board magnetic easel with 6 Penny Magnets in the boxes instead of dots. Say, **What is different about this Ten Frame?** Wait for a response.
- Say, **That is right. How many pennies are there?** Wait for a response.
- Say, **If I ask you how much money, I mean how many cents or how many pennies it is worth. So this is 6 cents. How much money is this?** Make sure the children use the word cents.
- Repeat by putting other numbers of pennies on the Ten Frame Mat until you are sure they have mastered the concept.
- *Note*: You can delay the introduction of the nickel until later lessons and begin with just pennies.
- Put the Nickel Magnet Strip up on the board above the Ten Frame Mat and say, **This is a nickel. A nickel is worth 5 pennies, like this row of pennies. We can trade these 5 pennies for a nickel.** Slide the 5 pennies off the top row of the Ten Frame Mat and replace with the Nickel Magnet Strip.
- Say, **If I ask you how much money, I mean how many cents or how many pennies it is worth. The nickel is worth 5 pennies so I count this 5** (circle the nickel), **6** (touch the penny) **cents.** Repeat.
- Say, **Let's try another one.** Leave up the nickel and put 2 pennies on the bottom row.
- Say, **How much money means how many pennies is it worth?** Touch the nickel and say **5,** then touch each penny as you say, **6, 7 cents.**
- Repeat with 8, 9, and 10 cents.

NUMBER OPERATIONS

★ Number Sentences with Sums More than 5

MATERIALS: White board magnetic easel, dry erase marker and eraser, two-colored dots, Vertical Flash Cards

Vertical Flash Card

- Say, **Sometimes we do not know the answer to a number sentence and we need to figure it out. When we add two numbers, it is easiest to start with the biggest number and then count up the other number. Let's try one.**
- Put the 6 + 2 Vertical Flash Card on the white board magnetic easel. Say, **This number sentence says, 6 plus 2 equals what? Let's put it up on the board.** Put up 6 red dots and 2 yellow dots, like the following.

- Say, **We can figure it out this way. I have 6** (circle the 6 dots) **and I am going to count on 2 more. 7, 8** (touch each of the bottom dots as you count). Say, **6,** (circle the 6 dots), **7, 8,** (touch each bottom dot as you count) Repeat.
- Say, **So 6 plus 2 equals 8.** Circle the 6 dots, then the 2 dots, then all the dots as you say each number.

Number Sense Interventions Lesson 19   113

- Say, **Let's write that down. 6 plus 2 equals 8.** Write the number sentence in the vertical format. Take down the Vertical Flash Card.
- Put up the 2 + 6 Vertical Flash Card and say, **This says 2 plus 6. Watch what I do!** Move the 2 dots over the 6 dots instead of under them.
- Say, **Did I take any dots away?** Wait for a response. **Did I add any dots?** Wait for a response. **So how many dots are here?** Circle all the dots and gesture to the sum 8 that you just wrote on the board.
- Say, **That is right! There are still 8 dots. I could count 2, 3, 4, 5, 6, 7, 8.** Gesture to the dots while counting. **But it takes so long! It is easier to do it this way: 6, 7, 8.** Gesture to the dots as you count.
- Say, **So 2 plus 6 equals 8. Let's write that down.** Write that next to the other Vertical Flash Card (6 + 2).
- Take the dots off but leave up the 2 Vertical Flash Cards. Say, **Let's try it without the dots.**
- Say, **Let's pretend I have 6 dots in this hand.** Show a closed fist. **And now I am going to count 2 more.** Hold up 2 fingers on the other hand.
- Say, **6 and 2 is 6, 7, 8.** Push your fist whenever you say, "6" and gesture with each of your 2 fingers as you count, "7, 8." Repeat.
- Say, **Now you try it with me. Make your hand with 6 pretend dots.** Make sure everyone is holding up a fist.
- Say, **Now put up 2 fingers for the 2 we are adding.** Make sure everyone is showing a fist and 2 fingers.
- Say, **Now we can count. Do what I do.** Using the previous gestures, say, **6, 7, 8.** Repeat.
- Take away the 2 Vertical Flash Cards. Put up the 7 + 3 Vertical Flash Card. Say, **Let's try another one. This one says, 7 plus 3. Let's pretend I have 7 dots in this hand.** Show a closed fist. **And now I am going to count 3 more.** Hold up 3 fingers on the other hand.
- Say, **7 and 3 is 7, 8, 9, 10.** Push your fist whenever you say, "7" and gesture with each of 2 fingers as you count, "8, 9, 10." Repeat.
- Say, **Now you try it with me. 7** (hold your fist) **plus 3** (hold up 3 fingers).
- Say, **Let's count. 7, 8, 9, 10.** Make sure everyone is using the correct gestures. Repeat.
- Repeat with 4 + 6 and 2 + 7 Vertical Flash Cards. Make sure you put the largest addend "in your hand."

Vertical Flash Card

Vertical Flash Card

STORY PROBLEMS

MATERIALS: Lesson 19 Activity Sheets, pencils, white board magnetic easel, dry erase marker and eraser

- Say, **Sometimes story problems have bigger numbers and it takes too long to draw the circles or sticks. Today I am going to tell some stories, but instead of making a drawing, I want you to write down the number sentence for the problem. Let's do one together.** This time write horizontal equations.

6 + 2 = 8

- Say, **Mike has 6 crayons. What number should I write down?** Wait for a response. Write a 6 on the white board magnetic easel.
- Say, **Jill has 2 crayons. What number should I write down?** Wait for a response. Write a 2 on the white board magnetic easel.
- Say, **How many crayons do they have altogether? That means we are adding so I will write a plus sign.** Put a plus sign between the 6 and 2.

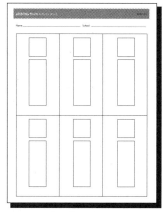
Lesson 19 Activity Sheet

- Say, **We know a new way to find 6 plus 2. 6 plus 2 is 6, 7, 8.** Use gestures for counting-on from previous activity.
- Say, **So, 6 crayons plus 2 crayons equals 8 crayons. Let's write equals 8.**
- Say, **Now you can try some with me!**
- Repeat with the following problems. Have the students write one problem in each box on the Lesson 19 Activity Sheets.
- Say, **Mike has 7 pencils. Jill gives him 3 more pencils. How many pencils does Mike have now?**
- Say, **Mike has 4 stickers. Jill gives him 4 more stickers. How many stickers does Mike have now?**
- Say, **Mike picked 2 flowers. Jill picked 8 flowers. How many flowers did they pick altogether?** (Be sure to start with the 8 to count up.)

Lesson 20

Learning Goals	Materials
Establish behavior boundaries	**COPY**
Count orally to 100	Hundreds Chart
Build 20 as 2 tens	Activity 20 Hundreds Chart
Count and sequence to 20	2 Decade Cards (10)
Number recognition 0–20	Teacher Ten Frame Mat
Before and after (1–10)	Lesson 20 Activity Sheet
Find $n + 1$, $n - 1$ for larger numbers	**GATHER**
Make numbers 1–10 on fingers	White board magnetic easel
Ten Frames: Nickel and pennies	20 interlocking blocks
Number operations on fingers: Counting-on	Dry erase marker and eraser
Use number sentences to solve story problems	Pencils without erasers and crayons
	PREPARE
	Put student names on Lesson 20 Activity Sheets.
	Number Recognition Cards (0–20); *See Chapter 1 for instructions on making Number Recognition Cards.*
	Vertical Flash Cards with sums of 6, 7, or 8; *See Chapter 1 for information on Vertical Flash Cards.*
	Penny Magnets for Ten Frame Mat; *See Chapter 1 for information on making Penny Magnets.*
	Nickel Magnet Strips for Ten Frame Mat; *See Chapter 1 for information on making Nickel Magnet Strips.*

ESTABLISHING BEHAVIOR BOUNDARIES

- Say, **Show me you are ready to learn.** Wait for proper posture. **Exactly! That shows me you are ready!**

Number Sense Interventions Lesson 20

COUNTING WARM-UP

MATERIALS: Hundreds Chart, white board magnetic easel

- Put up the Hundreds Chart to use for reference when needed.
- Say, **Today we are going to count higher numbers. I am going to start counting and then you keep going.**
- Say, **This time, you are each going to take a turn saying the next 3 numbers.**
- Let each child have 2 turns. Use the Hundreds Chart to correct errors.
- Do the following patterns: 25, 26, 27; 36, 37, 38; 47, 48, 49; 55, 56, 57; 66, 67, 68; 77, 78, 79; 85, 86, 87; 96, 97, 98.

1	2	3	4	5	6	7	8	9	10
11	12	13	14	15	16	17	18	19	20
21	22	23	24	25	26	27	28	29	30
31	32	33	34	35	36	37	38	39	40
41	42	43	44	45	46	47	48	49	50
51	52	53	54	55	56	57	58	59	60
61	62	63	64	65	66	67	68	69	70
71	72	73	74	75	76	77	78	79	80
81	82	83	84	85	86	87	88	89	90
91	92	93	94	95	96	97	98	99	100

Hundreds Chart

HUNDREDS CHART ACTIVITY

MATERIALS: Activity 20 Hundreds Chart, pencils

- Hand out the Activity 20 Hundreds Chart. Say, **Here is part of a Hundreds Chart, but 3 of the numbers are missing. I would like you to fill in the missing numbers. Please think carefully and do your own work.**
- Hand out pencils.

1	2	3	4	5	6	7	8	9	10
11	12	13		15	16	17	18	19	20
21	22	23	24	25		27	28	29	30
31	32	33	34	35	36	37	38		40

Activity 20 Hundreds Chart

MAGIC NUMBER ACTIVITIES (Magic Number is 20)

★ Making the Number 20

MATERIALS: Two 10 Decade Cards, 20 interlocking blocks

- Put down 20 unattached interlocking blocks. Say, **Here are some blocks. Let's count them together. You count on your fingers while I touch the blocks. Remember to use your inside voices. No shouting. Put your hands up so I can see them.**
- Count to 10, making a stick of 10 as you count. The children will have run out of fingers at 10. Hold up your fingers and say, **10 fingers! 10 blocks** (hold up the stick of 10 blocks).
- Say, **Let's keep counting.** Hold up the stick of 10 and say, "10". Touch each single block as you say, "11, 12, 13, 14, 15, 16, 17, 18, 19, 20."
- Say, **Let's show that on our fingers. 10** (hold up 10 fingers then put down hands and begin counting on fingers again), **11, 12, 13, 14, 15, 16, 17, 18, 19, 20.** Make sure children are counting correctly.
- Say, **We have used all our fingers again! 10 blocks, 10 fingers! Let's put these into another stick of 10.**
- Count the blocks as you put them into a stick.
- Say, **Now we have 2 sticks of 10. How many blocks are there? Let's count by tens. 10, 20.** Hold up each stick as you count. **20 blocks!**
- Hold up the 10 Decade Card and say, **What number is this?** Wait for a response. Say, **Yes, the number is 10.**
- Say, **I will put the number 10 right next to the 10 blocks.** Lay the stick down and put down the 10 Decade Card next to the stick.
- Say, **But we have another stick of 10! Here is the number 10 again.** Put the other 10 Decade Card next to the second stick of 10.

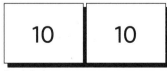

Decade Card

Decade Card

- Say, **Now watch what I do.** Put one 10 Decade Card over the other 10 Decade Card while saying, **10 and 10 more is 20. One 10 and one 10 is 2 tens. We call 2 tens, 20.** Take away the 2 tens and put down the 20 Decade Card.
- **This is the number 20. See the 2? That means 2 tens! What number is this?** Wait for a response.

★ Sequencing and Number Recognition 0-20

MATERIALS: Number Recognition Cards 0-20

Number Recognition Cards

- Say, **Say the numbers with me as I put them down. Do not go ahead of me. Ready? 1, 2, 3 ... 20.** Put the cards in 2 rows as if building a Hundreds Chart.
- Hold up the Number Recognition Card 20 and say, **What number is this?** Wait for a response.
- Say, **That is right. This is the number 20. Our Magic Number today is 20.**
- Say, **We are going to put the 20 here, under the 10. 20 is 10** (run your finger along the number list) **and 10 more** (point to the Number Recognition Card 20). Repeat.
- Say, **Let's play our game!**
- Go around the group, showing each child a different number from 0 to 20 (not in order). Make sure all children can see the number. Go around the group four times.

★ Before and After

MATERIALS: Number Recognition Cards 1-20

Number Recognition Cards

- Say, **I am going to show you a number, and I want you to tell me what number comes right after it. We will take turns, so do not call out if it is not your turn. Think carefully and tell me what number comes after.**
- Show the Number Recognition Cards 1-20 in random order. As you hold up a card say, **What number comes right after _____?**
- Give each child two turns. Repeat with "before" numbers.
- Error correction: If the child gives a wrong answer, then put the card on the table and say, **What number comes after/before _____** and point to the space after as in previous activities.

NUMBER LIST—NUMBER +1, −1

MATERIALS: Hundreds Chart, Number Recognition Cards 1-20

- Only show the Hundreds Chart if needed for error correction.
- Say, **Now we are going to play our plus one game! I have added more numbers today. You pick a card from this pile and we will add one to it.**
- Let the children continue to take turns drawing cards from the pile. Say, **How much is ___ + 1?**
- If the children hesitate or respond incorrectly, then use the Hundreds Chart to show plus one.
- Say, **Now we are going to play a different game! You pick a card from this pile and we will minus one from it.**
- Let the children continue to take turns drawing Number Recognition Cards from the pile. Say, **How much is ___ − 1?**
- If the children hesitate or respond incorrectly, then use the Hundreds Chart to show minus one.

Hundreds Chart

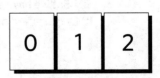

Number Recognition Cards

Number Sense Interventions Lesson 20

SUBITIZING QUANTITIES (1-10) ACTIVITIES

★ Finger Automaticity

MATERIALS: Number Recognition Cards 1-10

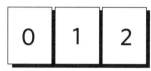
Number Recognition Cards

- Say, **Let's play our finger game. I will show you a card, and you show me the number on your fingers. Be as quick as you can!**
- Shuffle the Number Recognition Cards and show them one at a time. Go around the group three times. Correct errors.

★ Nickels and Pennies

MATERIALS: White board magnetic easel, Teacher Ten Frame Mat, Nickel Magnet Strip, 10 Penny Magnets

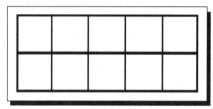
Ten Frame Mat

- Put the Teacher Ten Frame Mat on the white board magnetic easel with 6 pennies instead of dots. Say, **What is different about this Ten Frame?** Wait for a response.
- Say, **That is right. How many pennies are there?** Wait for a response.
- Put the Nickel Magnet Strip up on the white board magnetic easel and say, **This is a nickel. A nickel is worth 5 pennies, like this row of pennies. We can trade these 5 pennies for a nickel.** Slide the 5 pennies off the top row of the Ten Frame Mat and replace them with the Nickel Magnet Strip.
- Say, **If I ask you how much money, I mean how many cents or how many pennies it is worth. The nickel is worth 5 pennies, so I count like this: 5** (circle the nickel), **6** (touch the penny) **cents.** Repeat.
- Say, **Let's try another one.** Leave up the nickel and put 2 pennies on the bottom row.
- Say, **How much money means how many pennies is it worth?** Touch the nickel and say, **5,** then touch each penny as you say, **6, 7 cents.**
- Repeat with 8, 9, and 10 cents.

NUMBER OPERATIONS

★ Solving Number Sentences with Sums More than 5

MATERIALS: Vertical Flash Cards with sums of 6, 7, or 8

Vertical Flash Card

- Say, **Let's try adding with our fingers again. Remember, we start with the bigger number and then count up.**
- Put down the 4 + 2 Vertical Flash Card. Say, **Let's try this one. Let's hold 4 in this hand and make 2 on this hand.** Demonstrate.
- Say, **Let's count. 4** (gesture with a closed fist), **5, 6. 4 plus 2 equals 6.** Repeat. Make sure each child is showing a fist and 2 fingers and counting along.
- Repeat with remaining Vertical Flash Cards in random order.

SOLVING STORY PROBLEMS—MINUS

MATERIALS: Lesson 20 Activity Sheets, pencils, white board magnetic easel, dry erase marker and eraser

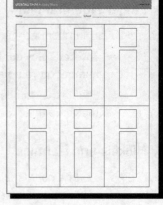
Lesson 20 Activity Sheet

- Hand out the Lesson 20 Activity Sheets and say, **Remember, sometimes story problems have bigger numbers and it takes too long to draw the circles or sticks. Today I am going to tell some stories, but instead of making a drawing, I want you to write down the number sentence for the problem like last time. But this time they will be take-away or minus problems. Let's do some together.**
- Say, **Mike has 10 pencils. What number should I write?** Wait for a response. **That is right. 10 pencils. Let's all write the number 10.**
- Say, **Jill takes away 2 of his pencils. What number should I write?** Wait for a response. **That is right. 2 pencils. Now you write the number.**
- Say, **The problem says takes away. What should I write here?** Point to the space between the numbers. Wait for a response.
- Say, **That is right! Take away is minus. So now our problem says 10 minus 2. How can we figure that out on our fingers?**
- Say, **Let's start with 10 fingers. Hold up 10 fingers like this.** Hold 10 fingers over the table close enough to touch the table with fingers when bent down.
- Say, **Watch me take away 2.** Touch the table with 2 fingers—pinkie and ring finger. **Now you try it!**
- Say, **How many pencils does Mike have now? 10 minus 2 is … Look at your fingers. How many are left? Tell me without counting.** Wait for a response.
- Say, **So, 10 pencils take away 2 pencils equals 8 pencils. Let's write 8 in the box.**
- Say, **Now you can try some with me!**
- Repeat with the following problems. Have them write one problem in each box on the Lesson 20 Activity Sheets.
- Say, **Mike has 10 crayons. Jill takes away 3 of his crayons. How many crayons does Mike have now?**
- Say, **Kisha picked 10 flowers but 4 of them died, so she threw them away. How many flowers does Kisha have now?**
- Say, **The balloon man had 5 balloons. 3 of them popped! How many balloons does the balloon man have now?**
- Say, **The farmer put 5 pigs into the pen. 2 pigs got out of the pen. How many pigs are in the pen now?**
- Say, **10 children went to the park. Only 5 of them got ice cream. How many children did not get ice cream?**

Lesson 21

Learning Goals

Establish behavior boundaries
Count orally to 100
Find missing numbers on Hundreds Chart
Build 21–29 as 2 tens and *n* units
Count and sequence to 29
Number recognition 0–29
Identify bigger/smaller number
Find *n* + 1, *n* − 1 with larger numbers
Make numbers 1–10 on fingers
Ten Frames: Nickel and pennies
Number operations on fingers: Counting-on
Use number sentences to solve story problems

Materials

COPY

Hundreds Chart
Cardinality Chart
Activity 21 Hundreds Chart
Decade Card (20)
Unit Cards (1–9)
Bigger/Smaller Cards
Lesson 21 Activity Sheet

GATHER

White board magnetic easel
21 interlocking blocks
Dry erase marker and eraser
Pencils without erasers and crayons

PREPARE

Put student names on Lesson 21 Activity Sheets.
Number Recognition Cards (0–29); *See Chapter 1 for instructions on making Number Recognition Cards.*
Vertical Flash Cards with sums of 7, 8, 9; *See Chapter 1 for information on Vertical Flash Cards.*
Penny Flash Cards; *See Chapter 1 for instructions on making Penny Flash Cards.*
Nickel and Penny Flash Cards; *See Chapter 1 for instructions on making Nickel and Penny Flash Cards.*

ESTABLISHING BEHAVIOR BOUNDARIES

- Say, **Show me you are ready to learn.** Wait for proper posture. **Exactly! That shows me you are ready!**

COUNTING WARM-UP

MATERIALS: Hundreds Chart, white board magnetic easel

- Put up the Hundreds Chart to use for reference when needed.
- Say, **Today we are going to count higher numbers. I am going to start counting and then you keep going.**
- Say, **This time, you are each going to take a turn saying the next 3 numbers.**
- Let each student have two turns. Use the Hundreds Chart to correct errors.
- Do the following patterns: 25, 26, 27; 36, 37, 38; 47, 48, 49; 55, 56, 57; 66, 67, 68; 77, 78, 79; 85, 86, 87; 96, 97, 98.

1	2	3	4	5	6	7	8	9	10
11	12	13	14	15	16	17	18	19	20
21	22	23	24	25	26	27	28	29	30
31	32	33	34	35	36	37	38	39	40
41	42	43	44	45	46	47	48	49	50
51	52	53	54	55	56	57	58	59	60
61	62	63	64	65	66	67	68	69	70
71	72	73	74	75	76	77	78	79	80
81	82	83	84	85	86	87	88	89	90
91	92	93	94	95	96	97	98	99	100

Hundreds Chart

HUNDREDS CHART ACTIVITY

MATERIALS: Activity 21 Hundreds Chart, pencils

- Hand out the Activity 21 Hundreds Chart. Say, **Here is part of a Hundreds Chart, but 3 of the numbers are missing. I would like you to fill in the missing numbers. Please think carefully and do your own work.**
- Hand out pencils.

1	2	3	4	5	6	7	8	9	10
11		13	14	15	16	17	18	19	20
21	22	23	24	25	26	27		29	30
31	32	33	34		36	37	38	39	40

Activity 21 Hundreds Chart

MAGIC NUMBER ACTIVITIES (Magic Number is 21)

★ Making the Number 21

MATERIALS: 20 Decade Card, Unit Cards 1–9, 21 interlocking blocks (2 sticks of 10 plus 1 block)

- Put down 21 interlocking blocks: 2 sticks of 10 and one individual block. Say, **We have 2 sticks of 10. How many blocks are there? Let's count by tens. 10, 20.** Hold up each stick as you count. **20 blocks!**
- Say, **20 and 1 more is 21.** Touch the 1 block as you say, "1 more."
- Hold up the 20 Decade Card and say, **What number is this?** Wait for a response. Say, **Yes, the number is 20.**
- Say, **I will put the number 20 right next to the 20 blocks.** Lay the 2 sticks down and put down the 20 Decade Card next to the sticks.
- Hold up the 1 Unit Card and say, **What number is this?** Wait for a response. **That is right. This is the number 1. I will put the number 1 right next to the 1 block.**
- Say, **Now watch what I do.** Put the 1 Unit Card over the 0 on the 20 Decade Card while saying, **20 and 1 more is 21. This is the number 21. What number is this?** Wait for a response.
- Say, **That is right. This is the number 21. See the 20 hiding under the 1** (lift up the 1 Unit Card)? **21 is 20 and 1 more** (show the 20 then add the 1 on top as you are saying, "and 1 more").
- Then say, **20** (circle the sticks of 10) **and 1 more** (circle the 1) **is 21.** Repeat.
- Take away the 1 Unit Card and add another block to the 1 block and say, **20 and 2 more is 22.**
- Put the 2 Unit Card on the 20 and say, **20 and 2 more is 22. What number is this?** Wait for a response.
- Repeat with 3–9 Unit Cards and blocks.

Decade Card

1

Unit Card

★ Sequencing and Number Recognition 1–29

MATERIALS: Number Recognition Cards 1–29

- Say, **Say the numbers with me as I put them down. Do not go ahead of me. Ready? 1, 2, 3, 4, 5 … 17, 18, 19, 20.** Put the cards in 2 rows as if building a Hundreds Chart.
- Hold up the Number Recognition Card 21 and say, **What number is this?** Wait for a response.
- Say, **That is right. This is the number 21. Our Magic Number today is 21.**
- Say, **We are going to put the 21 here, under the 11. 21 is 20** (circle the 20 cards) **and 1 more.**
- Say, **Let's put down more. 22, 23, 24 … 29.** Put down the specific Number Recognition Cards in the row as you say the number.
- Pick up Number Recognition cards 1–10 and put them aside. Pick up the remaining cards for the game. Say, **Let's play our game! Remember, our Magic Number is 21.**
- Go around the group, showing each child a different number from 11–29 (not in order). Make sure all children can see the number. Go around the group four times.

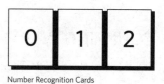

Number Recognition Cards

Number Sense Interventions Lesson 21   121

BIGGER/SMALLER

MATERIALS: Cardinality Chart, Bigger/Smaller Cards, white board magnetic easel

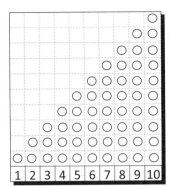
Cardinality Chart [number chart]

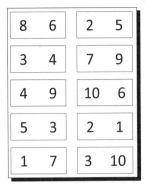
Bigger/Smaller Cards

- Show the Cardinality Chart only for error correction.
- Say, **I will show you a card with 2 numbers on it, and you tell me which number is bigger.**
- Go around the group twice, showing the Bigger/Smaller Cards in random order, saying, **Which number is bigger or more?**
- Say, **Now we are going to find smaller numbers. I will show you a card with 2 numbers on it, and you tell me which number is smaller.**
- Go around the group twice, showing the Bigger/Smaller Cards in random order, saying, **Which number is smaller or less?**

NUMBER LIST—NUMBER +1, −1

MATERIALS: Hundreds Chart, Number Recognition Cards 1–20

- Only show the Hundreds Chart if needed for error correction.
- Say, **Now we are going to play our plus one game! You pick a card from this pile and we will add one to it.**
- Let the children continue to take turns drawing Number Recognition Cards from the pile. Say, **How much is ___ + 1?**
- If the children hesitate or respond incorrectly, then use the Hundreds Chart to show plus one.
- Say, **Now we are going to play a different game! You pick a card from this pile and we will minus one from it.**
- Let the children continue to take turns drawing cards from the pile. Say, **How much is ___ − 1?**
- If the children hesitate or respond incorrectly, then use the Hundreds Chart to show minus one.

1	2	3	4	5	6	7	8	9	10
11	12	13	14	15	16	17	18	19	20
21	22	23	24	25	26	27	28	29	30
31	32	33	34	35	36	37	38	39	40
41	42	43	44	45	46	47	48	49	50
51	52	53	54	55	56	57	58	59	60
61	62	63	64	65	66	67	68	69	70
71	72	73	74	75	76	77	78	79	80
81	82	83	84	85	86	87	88	89	90
91	92	93	94	95	96	97	98	99	100

Hundreds Chart

Number Recognition Cards

SUBITIZING QUANTITIES (1-10) ACTIVITIES

★ Finger Automaticity

MATERIALS: Number Recognition Cards 1-10

- Say, **Let's play our finger game. I will show you a card, and you show me the number on your fingers. Be as quick as you can!**
- Shuffle the Number Recognition Cards and show them one at a time. Go around the group three times. Correct errors.

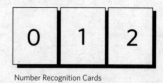
Number Recognition Cards

★ Nickels and Pennies

MATERIALS: Penny Flash Cards, Nickel and Penny Flash Cards

- Say, **Now I am going to hold up a card with some pennies on it, and I want you to tell me how many pennies are on the card. Try to tell me as fast as you can without counting. I will point to you for your turn, okay? Everyone else, say it in your mind. Remember to say, "cents" each time.**
- Go through the Penny Flash Cards in random order. Go around the group three times. Correct errors.
- Say, **Now I am going to hold up a card with a nickel and some pennies on it, and I want you to tell me how much money is on the card. Remember, a nickel is worth 5 pennies or 5 cents. Hold up 5 fingers. Try to tell me as fast as you can without counting. I will point to you for your turn, okay? Everyone else, say it in your mind.**
- Go through the Nickel and Penny Flash Cards in random order. Go around the group three times. Correct errors.

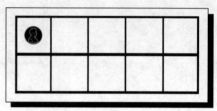
Penny Flash Cards

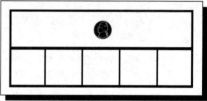
Nickel Flash Cards

NUMBER OPERATIONS

★ Solving Equations with Sums More than 5

MATERIALS: Vertical Flash Cards with sums of 7, 8, or 9

- Say, **Let's try adding with our fingers again. Remember, we start with the bigger number and then count up.**
- Put down the 5 + 2 Vertical Flash Card. Say, **Let's try this one. Let's hold 5 in this hand and make 2 on this hand.** Demonstrate.
- Say, **Let's count. 5, 6, 7. 5 plus 2 equals 7.** Repeat. Make sure each child is showing a fist and 2 fingers and counting along.
- Repeat with remaining Vertical Flash Cards in random order.

Vertical Flash Card

Number Sense Interventions Lesson 21   123

SOLVING STORY PROBLEMS

MATERIALS: Lesson 21 Activity Sheets, pencils, white board magnetic easel, dry erase marker and eraser

- Hand out the Lesson 21 Activity Sheets and say, **Remember, sometimes story problems have bigger numbers and it takes too long to draw the circles or sticks. Today I am going to tell some stories, but instead of making a drawing, I want you to write down the number sentence for the problem like last time. This time some will be plus and some will be take-away or minus problems. Let's do some together.**
- Read the following problems. Encourage children to work independently. After each story is completed, repeat the story and model on the white board magnetic easel. Have children check their answers but not erase them.
- Say, **Mike has 3 crayons. Jill has 2 crayons. How many crayons do they have altogether?**
- Say, **6 frogs were sitting on a log. 1 more frog jumps onto the log. How many frogs are on the log now?**
- Say, **Lizzy found 4 pennies on her way to school, and she found 2 more pennies on her way home. How many pennies did she find altogether?**
- Say, **Now these problems are a little different. Listen carefully.** Do not say, "we are doing minus."
- Say, **Mike has 3 pencils. He dropped 1 on the way to school. How many pencils does Mike have now?**
- Say, **Suzanne has 6 crayons. She broke 2 of her crayons, so she threw them away. How many crayons does Suzanne have now?**
- Say, **Aniyah brought 5 erasers to school. She gave 3 to her friends. How many erasers does Aniyah have now?**

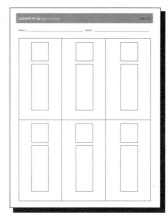

Lesson 21 Activity Sheet

Lesson 22

Learning Goals	Materials
Establish behavior boundaries	**COPY**
Count to 100 orally	Hundreds Chart
Find missing numbers on Hundreds Chart	Activity 22 Hundreds Chart
Build numbers 1–99 as *n* sets of 10 and *n* units	Decade Cards (all)
Count and sequence to 30	Unit Cards (all)
Number recognition 0–30	Lesson 22 Activity Sheet
Before and after (1–20)	**GATHER**
Find $n + 1$, $n - 1$ with larger numbers	White board magnetic easel
Make numbers 1–10 on fingers	30 interlocking blocks (3 sticks of 10)
Recognize money on Ten Frames	Dry erase marker and eraser
Number operations on fingers: Counting-on	Pencils without erasers and crayons
Use number sentences to solve story problems	**PREPARE**
Solve written combinations	Put student names on Lesson 22 Activity Sheets.
	Number Recognition Cards (0–30); *See Chapter 1 for instructions on making Number Recognition Cards.*
	Penny Flash Cards; *See Chapter 1 for instructions on making Penny Flash Cards.*
	Nickel and Penny Flash Cards; *See Chapter 1 for instructions on making Nickel and Penny Flash Cards.*
	Vertical Flash Cards with sums of 8, 9, or 10; *See Chapter 1 for information on Vertical Flash Cards.*

ESTABLISHING BEHAVIOR BOUNDARIES

- Say, **Show me you are ready to learn.** Wait for proper posture. **Exactly! That shows me you are ready!**

COUNTING WARM-UP

MATERIALS: Hundreds Chart, white board magnetic easel

- Put up the Hundreds Chart to use for reference when needed.
- Say, **Today we are going to count higher numbers. I am going to start counting and then you keep going.**
- Say, **This time, you are each going to take a turn saying the next 3 numbers.**
- Let them each have 2 turns. Use the Hundreds Chart to correct errors.
- Do the following patterns: 25, 26, 27; 36, 37, 38; 47, 48, 49; 55, 56, 57; 66, 67, 68; 77, 78, 79; 85, 86, 87; 96, 97, 98.

Hundreds Chart

Number Sense Interventions Lesson 22   125

HUNDREDS CHART ACTIVITY

MATERIALS: Activity 22 Hundreds Chart, pencils

- Hand out the Activity 22 Hundreds Chart. Say, **Here is part of a Hundreds Chart, but 3 of the numbers are missing. I would like you to fill in the missing numbers. Please think carefully and do your own work.**
- Hand out pencils.

1	2	3	4	5	6	7	8	9	10
11	12	13	14	15	16	17	18		20
21	22		24	25	26	27	28	29	30
31	32	33	34	35		37	38	39	40

Activity 22 Hundreds Chart

MAGIC NUMBER ACTIVITIES (1-100)

★ Making Numbers 1-100

MATERIALS: All Decade Cards including 100, Unit Cards 1-9, 3 sticks of 10 interlocking blocks

Note: Line up the Decade Cards in a horizontal line as you do this activity.

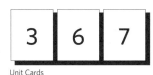
Decade Cards

- Say, **Let's count by tens. I will put down the numbers as you say them.** Put down each Decade Card as the students say, "10, 20, 30 … 100."
- Point to the non-zero numeral in each decade as you say, **10 is 1** (point to the 1 and hold up 1 stick) **stick of 10, 20 is 2** (point to the 2 and hold up 2 sticks) **sticks of 10, 30 is 3** (point to the 3 and hold up 3 sticks) **sticks of 10,** and so forth until 90 is 9 sticks of 10. Only model up to 3 sticks of 10 with interlocking blocks.

Unit Cards

- Say, **Today we are going to use all these cards to build more numbers! I will make a number and you tell me what it is. If I put 3 on 20, I make 23.** Circle the 20 Decade Card when you say, "20" and put down the 3 Unit Card when you say, "3." **What number is this?** Wait for a response.
- Take the 3 Unit Card and place it on each of the Decade Cards, one at a time, to make all the two-digit numbers with 3 in the unit place (i.e., 33, 43, 53, 63, 73, 83, 93), using the previous protocol. End with putting the 3 Unit Card on the 10 Decade Card. Go back over the numbers in which the students had trouble.

Note: Children often have trouble switching from the pattern of 23, 33, 43, …93 to then 13. If your children are having this problem, give them several opportunities to perform the switch by moving the 3 Unit Card between the 10 Decade Card and any other Decade Card before moving on to the next activity.

- Repeat the activity with the 6 Unit Card.
- Say, **We are going to play a game. I will build a number and then you tell me what the number is. We will take turns.**
- Put the 7 Unit Card on each of the Decade Cards beginning with 10. Go around the group until you reach 97.

★ Sequencing and Number Recognition 0-30

MATERIALS: Number Recognition Cards 0-30

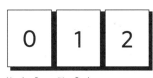
Number Recognition Cards

Note: This activity builds the beginning of a Hundreds Chart by lining up the Number Recognition Cards in rows of 10.

- Say, **Say the numbers with me as I put them down. Do not go ahead of me. Ready? 1, 2, 3 … 30.** Put the cards in 3 rows as if building a Hundreds Chart.
- Hold up the Number Recognition Card 30 and say, **What number is this?** Wait for a response.
- Say, **That is right. This is the number 30. Our Magic Number today is 30. 3 sticks of 10 are 30 blocks.**

- Say, **We are going to put the 30 here, under the 20.** Place the Number Recognition Card 30 under the Number Recognition Card 20.
- Say, **Let's play our game!**
- Go around the group, showing each child a different number from 0 to 30 (not in order). Make sure all children can see the number. Go around the group four times.

★ Before and After

MATERIALS: Number Recognition Cards 1–20

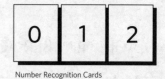
Number Recognition Cards

- Say, **I am going to show you a number, and I want you to tell me what number comes right after it. We will take turns, so do not call out if it is not your turn. Think carefully and tell me what number comes after.**
- Show Number Recognition Cards 1–20 in random order. As you hold up a card say, **What number comes right after _____?**
- Give each child two turns. Repeat with "before" numbers.
- Error correction: If the child gives a wrong answer, then put the card on the table and say, **What number comes after/before ____** and point to the space after as in previous activities.

NUMBER LIST—NUMBER +1, –1

MATERIALS: Hundreds Chart, Number Recognition Cards 1–20

- Only show the Hundreds Chart if needed for error correction.
- Say, **Now we are going to play our plus one game! You pick a card from this pile and we will add 1 to it.**
- Let the children continue to take turns drawing Unit Cards from the pile. Say, **How much is ___ + 1?**
- If the children hesitate or respond incorrectly, then use the Hundreds Chart to show plus one.
- Say, **Now we are going to play a different game! You pick a card from this pile and we will minus one from it.**
- Let the children continue to take turns drawing Unit Cards from the pile. Say, **How much is ___ – 1?**
- If the children hesitate or respond incorrectly, then use the Hundreds Chart to show minus one.

Hundreds Chart

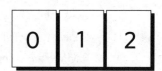
Number Recognition Cards

SUBITIZING QUANTITIES (1–10) ACTIVITIES

★ Finger Automaticity

MATERIALS: Number Recognition Cards 1–10

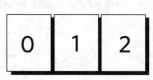
Number Recognition Cards

- Say, **Let's play our finger game. I will show you a card, and you show me the number on your fingers. Be as quick as you can!**
- Shuffle the Number Recognition Cards and show them one at a time. Go around the group three times. Correct errors.

★ Nickels and Pennies

MATERIALS: Penny Flash Cards, Nickel and Penny Flash Cards

- Say, **Now I am going to hold up a card with some pennies on it, and I want you to tell me how many pennies are on the card. Try to tell me as fast as you can without counting. I will point to you for your turn, okay? Everyone else, say it in your mind. Remember to say, "cents" each time.**

- Go through the Penny Flash Cards in random order. Go around the group three times. Correct errors.

- Say, **Now I am going to hold up a card with a nickel and some pennies on it, and I want you to tell me how much money is on the card. Remember, a nickel is worth 5 pennies or 5 cents. Hold up 5 fingers. Try to tell me as fast as you can without counting. I will point to you for your turn, okay? Everyone else, say it in your mind.**

- Go through the Nickel and Penny Flash Cards in random order. Go around the group three times. Correct errors.

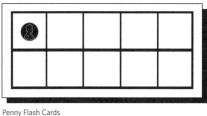
Penny Flash Cards

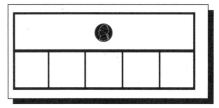
Nickel Flash Cards

NUMBER OPERATIONS

★ Solving Number Sentences with Sums More than 5

MATERIALS: Vertical Flash Cards with sums of 8, 9, or 10

- Say, **Let's try adding with our fingers again. Remember, we start with the bigger number and then count up.**

- Put down the 7 + 2 Vertical Flash Card. Say, **Let's try this one. Let's hold 7 in this hand and make 2 on this hand.** Demonstrate.

- Say, **Let's count. 7, 8, 9. 7 plus 2 equals 9.** Repeat. Make sure each child is showing a fist and 2 fingers and counting along.

- Repeat with remaining Vertical Flash Cards in random order.

Vertical Flash Card

SOLVING STORY PROBLEMS

MATERIALS: Lesson 22 Activity Sheets, pencils, white board magnetic easel, dry erase marker and eraser

- Hand out the Lesson 22 Activity Sheets and say, **I am going to tell some stories, and I want you to write down the number sentence for the problems. Do not draw the problem. Just write the number sentence. Listen carefully to decide if it is plus or minus. If you know the answer, then just write it down. If you are not sure, then use your fingers to figure it out.**

- Read the following problems one at a time. Encourage children to work independently. After each story is completed, repeat the story and model on the white board magnetic easel. Have children check their answers but not erase them.

- Say, **Joe had 6 pennies. His mother gave him 3 more. How many pennies does Joe have now?**

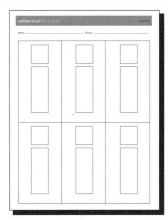
Lesson 22 Activity Sheet

- Say, **Sam took 4 steps forward. Then he took 4 more steps forward. How many steps forward did he take altogether?**
- Say, **Tamika picked up 5 stones on the sidewalk. Then she picked up 4 more on the playground. How many stones did she pick up altogether?**
- Say, **Now these problems are a little different. Listen carefully.** Do not say, "we are doing minus."
- Say, **Mike has 8 pennies. He dropped 2 on the way to school. How many pennies does Mike have now?**
- Say, **Kiera drew 6 pictures. She threw 2 of her pictures away. How many pictures does Kiera have now?**
- Say, **Jude brought 10 cupcakes to school. He gave 8 to the teachers. How many cupcakes does Jude have now?**

Lesson 23

Learning Goals	Materials
Establish behavior boundaries	**COPY**
Count to 100 orally	Hundreds Chart
Find missing numbers on Hundreds Chart	Activity 23 Hundreds Chart
Build numbers 1–99 as n sets of 10 and n units	Decade Cards (all)
Count and sequence to 30	Unit Cards 1–9
Number recognition 0–29	Cardinality Chart
Identify bigger/smaller number	Bigger/Smaller Cards
Find $n + 1$, $n - 1$ with larger numbers	Teacher Ten Frame Mat
Make numbers 1–10 on fingers	Lesson 23 Activity Sheet
Recognize money on Ten Frames	**GATHER**
Number operations on fingers: Counting-on	White board magnetic easel
Use number sentences to solve story problems	30 interlocking blocks made into 3 sticks of 10
Solve written combinations	Dry eraser marker and eraser
	Pencils without erasers and crayons
	PREPARE
	Put student names on Lesson 23 Activity Sheets.
	Number Recognition Cards (0–30); *See Chapter 1 for instructions on making Number Recognition Cards.*
	Penny Flash Cards; *See Chapter 1 for instructions on making Penny Flash Cards.*
	10 magnetic pennies
	Nickel and Penny Flash Cards; *See Chapter 1 for instructions on making Nickel and Penny Flash Cards.*
	Magnetic Dime
	Vertical Flash Cards with sums of 6–10; *See Chapter 1 for more information on Vertical Flash Cards.*

ESTABLISHING BEHAVIOR BOUNDARIES

- Say, **Show me you are ready to learn.** Wait for proper posture. **Exactly! That shows me you are ready!**

Number Sense Interventions Lesson 23   129

COUNTING WARM-UP

MATERIALS: Hundreds Chart, white board magnetic easel

- Put up the Hundreds Chart to use for reference when needed.
- Say, **Today we are going to count higher numbers. I am going to start counting and then you keep going.**
- Say, **This time, you are each going to take a turn saying the next 3 numbers.**
- Let students each have 2 turns. Use the Hundreds Chart to correct errors.
- Do the following patterns: 25, 26, 27; 36, 37, 38; 47, 48, 49; 55, 56, 57; 66, 67, 68; 77, 78, 79; 85, 86, 87; 96, 97, 98.

Hundreds Chart

HUNDREDS CHART ACTIVITY

MATERIALS: Activity 23 Hundreds Chart, pencils

- Hand out the Activity 23 Hundreds Chart. Say, **Here is part of a Hundreds Chart, but several of the numbers are missing. I would like you to fill in the missing numbers. Please think carefully and do your own work.**
- Hand out pencils.

Activity 23 Hundreds Chart

MAGIC NUMBER ACTIVITIES (1-100)

★ Making Numbers 1-100

MATERIALS: All Decade Cards including 100, all Unit Cards, 3 sticks of 10 interlocking blocks

Decade Cards

Unit Cards

Note: Line up the Decade Cards in a horizontal line as you do this activity.

- Say, **Let's count by tens. I will put down the numbers as you say them.** Put down each Decade Card as the students say, "10, 20, 30 … 100."
- Point to the non-zero numeral in each decade as you say, **10 is 1** (point to the 1 and hold up 1 stick) **stick of 10, 20 is 2** (point to the 2 and hold up 2 sticks) **sticks of 10, 30 is 3** (point to the 3 and hold up 3 sticks) **sticks of 10,** and so forth until 90 is 9 sticks of 10. Only model up to 3 sticks of ten with interlocking blocks.
- Say, **Today we are going to use all these cards to build more numbers! I will make a number and you tell me what it is. If I put 4 on 20, I make 24.** Circle the 20 Decade Card when you say, "20" and put down the 4 Unit Card when you say, "4." **What number is this?** Wait for a response.
- Place the 4 Unit Card on each of the Decade Cards one at a time to make all the two-digit numbers with 4 in the unit place (i.e., 44, 54, 64, 74, 84, 94) using the previous protocol. Go back over the numbers in which the students had trouble. Finally, place the 4 Unit Card on the 10 Decade Card.
- Repeat with the 5 Unit Card.

- Say, **We are going to play a game. I will build a number and then you tell me what the number is. We will take turns.**
- Put the 2 Unit Card on successive decades beginning with 10. Go around the group until you reach 92.

★ Sequencing and Number Recognition 0-30

MATERIALS: Number Recognition Cards 0-30

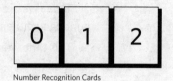
Number Recognition Cards

Note: This activity builds the beginning of a Hundreds Chart by lining up the Number Recognition Cards in rows of 10.

- Say, **Say the numbers with me as I put them down. Do not go ahead of me. Ready? 1, 2, 3 … 30.** Put the cards in three rows as if building a Hundreds Chart.
- Hold up the Number Recognition Card 30 and say, **What number is this?** Wait for a response.
- Say, **That is right. This is the number 30. Our Magic Number today is 30. 3 sticks of 10 are 30 blocks.**
- Say, **We are going to put the 30 here, under the 20.** Place the Number Recognition Card 30 under the Number Recognition Card 20.
- Say, **Let's play our game!**
- Go around the group, showing each child a different number from 0 to 30 (not in order). Make sure all children can see the number. Go around the group four times.

BIGGER/SMALLER

MATERIALS: Cardinality Chart, Bigger/Smaller Cards, white board magnetic easel

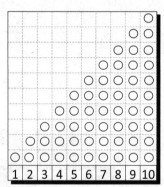
Cardinality Chart [number chart]

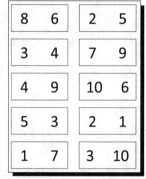
Bigger/Smaller Cards

- Show the Cardinality Chart only for error correction.
- Say, **I will show you a card with 2 numbers on it, and you tell me which number is bigger.**
- Go around the group twice, showing the Bigger/Smaller Cards in random order, saying, **Which number is bigger or more?**
- Say, **Now we are going to find smaller numbers. I will show you a card with 2 numbers on it, and you tell me which number is smaller.**
- Go around the group twice, showing the Bigger/Smaller Cards in random order, saying, **Which number is smaller or less?**

Number Sense Interventions Lesson 23   131

NUMBER LIST—NUMBER +1, −1

MATERIALS: Hundreds Chart, Number Recognition Cards 1-20

- Only show the Hundreds Chart if needed for error correction.
- Say, **Now we are going to play our plus one game! You pick a card from this pile and we will add one to it.**
- Let the children continue to take turns drawing Unit Cards from the pile. Say, **How much is ___ + 1?**
- If the children hesitate or respond incorrectly, then use the Hundreds Chart to show plus one.
- Say, **Now we are going to play a different game! You pick a card from this pile and we will minus one from it.**
- Let the children continue to take turns drawing Unit Cards from the pile. Say, **How much is ___ − 1?**
- If the children hesitate or respond incorrectly, then use the Hundreds Chart to show minus one.

Hundreds Chart

Number Recognition Cards

SUBITIZING QUANTITIES (1-10) ACTIVITIES

★ Finger Automaticity

MATERIALS: Number Recognition Cards 1-10

- Say, **Let's play our finger game. I will show you a card, and you show me the number on your fingers. Be as quick as you can!**
- Shuffle the Number Recognition Cards and show them one at a time. Go around the group three times. Correct errors.

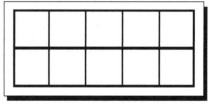
Number Recognition Cards

★ Nickels and Pennies

MATERIALS: White board magnetic easel, Teacher Ten Frame Mat, 10 magnetic pennies, dime, Penny Flash Cards, Nickel and Penny Flash Cards

- Put the Teacher Ten Frame Mat on the white board magnetic easel and cover it with pennies while saying, **I am going to fill the frame with pennies.**
- Say, **How many pennies? Tell me without counting?** Wait for a response.
- Hold up the dime and say, **Here is a dime. A dime is worth 10 pennies or 10 cents. We can trade these 10 pennies in for a dime!** Push the pennies off the frame and replace it with the dime. **A dime is worth 10 cents.**
- Take the dime off the Ten Frame and put up 6 pennies.
- Say, **How many pennies are here? Tell me without counting!** Wait for a response.
- Say, **How many more pennies do I need to make 10 cents?** Gesture to the 4 spaces. Wait for a response.
- Say, **How many more pennies do I need to make a dime?** Gesture to the 4 spaces. Wait for a response.
- Say, **Now I am going to hold up a card with some pennies on it, and I want you to tell me how much money is on the card. Then I want you to tell me how many more pennies I need to make a dime or 10 cents.**

Ten Frame Mat

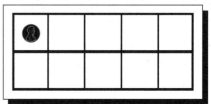
Penny Flash Cards

- Go through the Penny Flash Cards in random order. Go around the group three times. Correct errors.
- Say, **Now I am going to hold up a card with a nickel and some pennies on it, and I want you to tell me how much money is on the card. Then I want you to tell me how many more pennies I need to make a dime or 10 cents.** Go through the Nickel and Penny Flash Cards in random order. Go around the group three times. Correct errors.

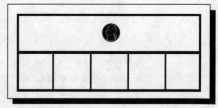
Nickel Flash Cards

NUMBER OPERATIONS

★ Solving Equations with Sums More than 5

MATERIALS: Vertical Flash Cards with sums of 6-10

Vertical Flash Card

- Say, **Let's try adding with our fingers again. Remember, we start with the bigger number and then count up.**
- Put down the 8 + 2 Vertical Flash Card. Say, **Let's try this one. Let's hold 8 in this hand and make 2 on this hand.** Demonstrate.
- Say, **Let's count. 8, 9, 10. 8 plus 2 equals 10.** Repeat. Make sure each child is showing a fist and 2 fingers and counting along.
- Repeat with remaining flash cards in random order.

SOLVING STORY PROBLEMS

MATERIALS: Lesson 23 Activity Sheets, pencils, white board magnetic easel, dry erase marker and eraser

- Hand out the Lesson 23 Activity Sheets and say, **I am going to tell some stories, and I want you to write down the number sentence for the problems. Do not draw the problem, just write the number sentence. Listen carefully to decide if it is plus or minus. If you know the answer, then write it down. If you are not sure, use your fingers to figure it out.**
- Read the following problems. Encourage children to work independently. After each story is completed, repeat the story and model on the white board magnetic easel. Have children check their answers but not erase them.

Lesson 23 Activity Sheet

- Say, **There are some pigs at the farm. 7 are in the pen and 2 are outside the pen. How many pigs are there altogether?**
- Say, **There are 6 pigs in the pen. The farmer brings 4 more into the pen. How many pigs are in the pen now?**
- Say, **There is 1 pig on the road. 9 more pigs came to join him. How many pigs are on the road now?**
- Say, **Now these problems are a little different. Listen carefully.** Do not say, "we are doing minus."
- Say, **The farmer bought 8 pigs at the market. 3 ran away when he was putting them into the pen. How many pigs are still in the pen?**
- Say, **The farmer bought 9 pigs at the market. He sold 3 of the pigs to his neighbor. How many pigs did the farmer have left?**
- Say, **The farmer put 7 pigs in the pen. 1 got sick, so he put it in the barn to stay warm. How many pigs were left in the pen?**

Number Sense Interventions Lesson 24

Lesson 24

Learning Goals	Materials
Establish behavior boundaries	**COPY**
Count to 100 orally	Hundreds Chart
Find missing numbers on the Hundreds Chart	Activity 24 Hundreds Chart
Build numbers 1-99 as *n* sets of 10 and *n* units	Decade Cards (all)
Count and sequence to 30	Unit Cards 1-9
Number recognition 0-30	Teacher Ten Frame Mat
Before and after (1-10)	Lesson 24 Activity Sheet
Find *n* + 1, *n* − 1 with larger numbers	**GATHER**
Make numbers 1-10 on fingers	White board magnetic easel
Recognize money on Ten Frames	30 interlocking blocks made into 3 sticks of 10
Number operations on fingers: Counting-on	Magnetic dime
Use number problems to solve story problems	10 magnetic pennies
	Dry erase marker and eraser
	Pencils without erasers and crayons

PREPARE

Put student names on Lesson 24 Activity Sheets.

Number Recognition Cards (0-30); *See Chapter 1 for instructions on making Number Recognition Cards.*

Penny Flash Cards; *See Chapter 1 for instructions on making Penny Flash Cards.*

Nickel and Penny Flash Cards; *See Chapter 1 for instructions on making Nickel and Penny Flash Cards.*

Vertical Flash Cards with sums of 6-10; *See Chapter 1 for information on Vertical Flash Cards.*

ESTABLISHING BEHAVIOR BOUNDARIES

- Say, **Show me you are ready to learn.** Wait for proper postures. **Exactly! That shows me you are ready!**

COUNTING WARM-UP

MATERIALS: Hundreds Chart, white board magnetic easel

- Put up the Hundreds Chart to use for reference when needed.
- Say, **Today we are going to count higher numbers. I am going to start counting and then you keep going.**
- Say, **This time, you are each going to take a turn saying the next 3 numbers.**
- Let each student have 2 turns. Use the Hundreds Chart to correct errors.
- Do the following patterns: 25, 26, 27; 36, 37, 38; 47, 48, 49; 55, 56, 57; 66, 67, 68; 77, 78, 79; 85, 86, 87; 96, 97, 98.

1	2	3	4	5	6	7	8	9	10
11	12	13	14	15	16	17	18	19	20
21	22	23	24	25	26	27	28	29	30
31	32	33	34	35	36	37	38	39	40
41	42	43	44	45	46	47	48	49	50
51	52	53	54	55	56	57	58	59	60
61	62	63	64	65	66	67	68	69	70
71	72	73	74	75	76	77	78	79	80
81	82	83	84	85	86	87	88	89	90
91	92	93	94	95	96	97	98	99	100

Hundreds Chart

HUNDREDS CHART ACTIVITY

MATERIALS: Activity 24 Hundreds Chart, pencils

- Hand out the Activity 24 Hundreds Chart. Say, **Here is part of a Hundreds Chart, but several of the numbers are missing. I would like you to fill in the missing numbers. Please think carefully and do your own work.**
- Hand out pencils.

Activity 24 Hundreds Chart

MAGIC NUMBER ACTIVITIES (1–100)

★ Making Numbers 1–100

MATERIALS: All Decade Cards including 100, Unit Cards 1–9, 3 sticks of 10 interlocking blocks

Decade Cards

Unit Card

Note: Line up the Decade Cards in a horizontal line as you do this activity.

- Say, **Let's count by tens. I will put down the numbers as you say them.** Put down each Decade Card as the students say, "10, 20, 30 … 100."
- Point to the non-zero numeral in each decade as you say, **10 is 1** (point to the 1 and hold up 1 stick) **stick of 10, 20 is 2** (point to the 2 and hold up 2 sticks) **sticks of 10, 30 is 3** (point to the 3 and hold up 3 sticks) **sticks of 10,** and so forth until 90 is nine sticks of 10.
- Say, **Today we are going to use all these cards to build more numbers! I will make a number and you tell me what it is. If I put 8 on 20, I make 28.** Circle the 20 Decade Card when you say, "20" and put down the 8 Unit Card when you say, "8." **What number is this?** Wait for a response.
- Place the 8 Unit Card on each of the Decade Cards to make all the two-digit numbers with 8 (i.e., 38, 48, 58, 68, 78, 88, 98), using the previous protocol. Go back over the numbers in which the students had trouble.
- Say, **Now we will play another game. I am going to say the number, and I want you to make the number.** Put down all the Unit Cards from which the children can choose. Call on them one at a time to make one of the following numbers by putting the appropriate Unit Card on the appropriate Decade Card. Use the numbers: 64, 35, 21, 83, 18, 59, 42, 96.

★ Sequencing and Number Recognition 0–30

Number Recognition Cards

MATERIALS: Number Recognition Cards 0–30

- Say, **Say the numbers with me as I put them down. Do not go ahead of me. Ready? 1, 2, 3 … 30.** Put the Number Recognition Cards in 3 rows as if building a Hundreds Chart.
- Hold up the Number Recognition Card 30 and say, **What number is this?** Wait for a response.
- Say, **That is right. This is the number 30. Our Magic Number today is 30. 3 sticks of 10 are 30 blocks.**
- Say, **We are going to put the 30 here, under the 20.** Place the Number Recognition Card 30 under the Number Recognition Card 20.
- Say, **Let's play our game!**
- Go around the group, showing each child a different number from 0 to 30 (not in order). Make sure all children can see the number. Go around the group four times.

Number Sense Interventions Lesson 24   135

★ Before and After

MATERIALS: Number Recognition Cards 1-20

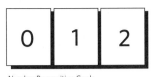
Number Recognition Cards

- Say, **I am going to show you a number, and I want you to tell me what number comes right after it. We will take turns, so do not call out if it is not your turn. Think carefully and tell me what number comes after.**
- Show Number Recognition Cards 1-20 in random order. As you hold up a card say, **What number comes right after _____?**
- Give each child two turns. Repeat with "before" numbers.
- Error correction: If the child gives a wrong answer, then put the card on the table and say, **What number comes after/before _____** and point to the space after, as in previous activities.

NUMBER LIST—NUMBER +1, −1

MATERIALS: Hundreds Chart, Number Recognition Cards 1-20

- Only show the Hundreds Chart if needed for error correction.
- Say, **Now we are going to play our plus one game! You pick a card from this pile and we will add one to it.**
- Let the children continue to take turns drawing cards from the pile. Say, **How much is ___ + 1?**
- If the children hesitate or respond incorrectly, then use the Hundreds Chart to show plus one.
- Say, **Now we are going to play a different game! You pick a card from this pile and we will minus one from it.**
- Let the children continue to take turns drawing cards from the pile. Say, **How much is ___ − 1?**
- If the children hesitate or respond incorrectly, then use the Hundreds Chart to show minus one.

1	2	3	4	5	6	7	8	9	10
11	12	13	14	15	16	17	18	19	20
21	22	23	24	25	26	27	28	29	30
31	32	33	34	35	36	37	38	39	40
41	42	43	44	45	46	47	48	49	50
51	52	53	54	55	56	57	58	59	60
61	62	63	64	65	66	67	68	69	70
71	72	73	74	75	76	77	78	79	80
81	82	83	84	85	86	87	88	89	90
91	92	93	94	95	96	97	98	99	100

Hundreds Chart

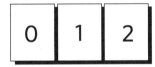
Number Recognition Cards

SUBITIZING QUANTITIES (1-10) ACTIVITIES

★ Finger Automaticity

MATERIALS: Number Recognition Cards 1-10

Number Recognition Cards

- Say, **Let's play our finger game. I will show you a card, and you show me the number on your fingers. Be as quick as you can!**
- Shuffle the Number Recognition Cards and show them one at a time. Go around the group three times. Correct errors.

★ Nickels, Pennies, and Dime

MATERIALS: White board magnetic easel, magnetic dime, 10 magnetic pennies, Teacher Ten Frame Mat, Penny Flash Cards, Nickel and Penny Flash Cards

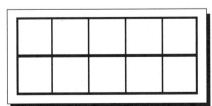
Ten Frame Mat

- Put up the Teacher Ten Frame Mat on the white board magnetic easel. Put 10 magnetic pennies on the Teacher Ten Frame Mat. Hold up the dime and say, **Here is a dime. A dime is worth 10 pennies or 10 cents. We can trade these 10 pennies in for a dime!** Push the pennies off the frame and replace with the dime. **A dime is worth 10 cents.**

- Take the dime off the Teacher Ten Frame Mat and put up 6 pennies.
- Say, **How many pennies are here? Tell me without counting!** Wait for a response.
- Say, **How many more pennies do I need to make 10 cents?** Gesture to the 4 spaces. Wait for a response.
- Say, **How many more pennies do I need to make a dime?** Gesture to the 4 spaces. Wait for a response.
- Say, **Now I am going to hold up a card with some pennies on it, and I want you to tell me how much money is on the card. Then I want you to tell me how many more pennies I need to make a dime or 10 cents.**
- Go through the Penny Flash Cards in random order. Go around the group three times. Correct errors.
- Say, **Now I am going to hold up a card with a nickel and some pennies on it, and I want you to tell me how much money is on the card. Then I want you to tell me how many more pennies I need to make a dime or 10 cents.** Go through the Nickel and Penny Flash Cards in random order. Go around the group three times. Correct errors.

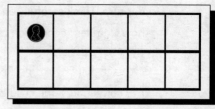

Penny Flash Cards

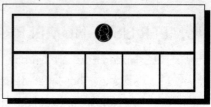

Nickel Flash Cards

NUMBER OPERATIONS

★ Solving Number Sentences with Sums More than 5

MATERIALS: Vertical Flash Cards with sums of 6–10

- Say, **Let's try adding with our fingers again. Remember, we start with the bigger number and then count up.**
- Put down the 8 + 2 Vertical Flash Card. Say, **Let's try this one. Let's hold 8 in this hand and make 2 on this hand.** Demonstrate.
- Say, **Let's count. 8, 9, 10. 8 plus 2 equals 10.** Repeat. Make sure each child is showing a fist and 2 fingers and counting along.
- Repeat with remaining Vertical Flash Cards in random order.

Vertical Flash Card

SOLVING STORY PROBLEMS

MATERIALS: Lesson 24 Activity Sheets, pencils, white board magnetic easel, dry erase marker with eraser

- Hand out the Lesson 24 Activity Sheets and say, **I am going to tell some stories, and I want you to write down the number sentence for the problems. Do not draw the problem, just write the number sentence. Listen carefully to decide if it is plus or minus. If you know the answer, then write it down. If you are not sure, then use your fingers to figure it out.**
- Read the following problems. Encourage children to work independently. After each story is completed, repeat the story and model on the white board magnetic easel. Have children check their answers but not erase them.
- Say, **Juan picked 5 apples. He picked 3 more apples. How many apples does Juan have now?**

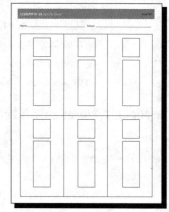

Lesson 24 Activity Sheet

Number Sense Interventions Extension Partner Activities   137

- Say, **There were 3 ducks in the pond and 7 in the grass. How many ducks were there altogether?**
- Say, **There were 2 cars in the parking lot. 6 more cars drove in. How many cars were in the parking lot?**
- Say, **Now these problems are a little different. Listen carefully.** Do not say, "we are doing minus."
- Say, **There were 6 ducks in the grass. 1 duck jumped into the pond. How many ducks were still in the grass?**
- Say, **The balloon man had 8 balloons. 5 of them popped! How many balloons does the balloon man have now?**
- Say, **Sam had 9 pennies. He gave 5 to his sister to buy a piece of candy. How many pennies did Sam have then?**

Extension Partner Activities

OVERVIEW

- Extension partner activities are intended for use after children have mastered partners through the 5 family. Each activity focuses on one family of partners (i.e., 6, 7, 8, 9, and 10 families). The teacher and children begin with the family number of dots—all turned to the red side. Next, one dot is turned to the yellow side and the children find the plus and minus number sentences associated with the combination of red and yellow dots as completed in previous lessons. For example, 5 red dots and 1 yellow dot match the number sentences 5 + 1 = 6, 1 + 5 = 6, 6 − 5 = 1, and 6 − 1 = 5. When one more dot is turned to the yellow side, there are now 4 red and 2 yellow dots; therefore, the matching number sentences are 4 + 2 = 6, 2 + 4 = 6, 6 − 4 = 2, and 6 − 2 = 4. One last partner, 3 red and 3 yellow dots, matches with the number sentences 3 + 3 = 6 and 6 − 3 = 3.

PARTNERS OF 6 ACTIVITIES

MATERIALS: Student Ten Frame Trays with 6 two-colored magnetic dots; white board magnetic easel; Teacher Ten Frame Mat and 6 two-colored magnetic dots; Number Sentence Cards: 5 + 1 = 6, 1 + 5 = 6, 6 − 5 = 1, 6 − 1 = 5, 4 + 2 = 6, 2 + 4 = 6, 6 − 4 = 2, 6 − 2 = 4, 3 + 3 = 6, 6 − 3 = 3

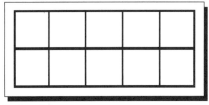
Ten Frame Mat

- Put the Teacher Ten Frame Mat on the white board magnetic easel. Add 6 red dots and say, **Here are 6 red dots. Let's see how many ways we can split 6 into two parts.**
- Turn over the bottom dot. **We already know that 5 plus 1 equals 6.** Gesture to the dots. **Let's show that on our fingers. 5 plus 1 equals 6.** Show on fingers. Make sure all the children show it correctly.

- Put down the plus cards: 5 + 1 = 6, 1 + 5 = 6, 4 + 2 = 6, 2 + 4 = 6, and 3 + 3 = 6. Say, **Which card shows the number sentence 5 plus 1 equals 6?** Call on someone to choose the correct card. If he or she chooses the incorrect card, then hold the card up to the dots to compare. Allow him or her to choose another card.

- Point to the yellow dot and say, **I can put the yellow dot in the sentence first. Then I say 1 plus 5 equals 6.** Gesture to the dots in that order. **Which card says 1 plus 5 equals 6?** Call on someone to choose the correct card.

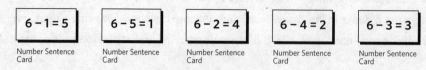

- Say, **Now let's find the number sentences that are minus.** Push the plus cards to the side and put down the minus cards: 6 − 1 = 5, 6 − 5 = 1, 6 − 2 = 4, 6 − 4 = 2, and 6 − 3 = 3.
- Say, **First, I will cover the yellow dot.** Cover the 1 yellow dot.
- Say, **Which number sentence am I showing? 6 minus ____.** Gesture covering the 1 yellow dot. Call on someone to choose the correct card. Remove cards from the pile each time.
- Say, **Now I will cover the red dots.** Cover the 5 red dots.
- Say, **Which number sentence am I showing? 6 minus ____.** Gesture covering the 5 red dots. Call on someone to choose the correct card. Remove cards from the pile each time.
- Say, **Let's try another one! I am going to give you each a frame and 6 dots.** Pass out the Student Ten Frame Trays with 6 two-colored magnetic dots (4 red dots and 2 yellow dots). The dots should look like this:

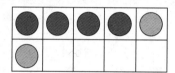

- Change your dots to match the children's dots. Say, **What number sentence does this say?**
- Repeat the previous protocol with 4 red and 2 yellow dots. Have children copy you. Be sure to point out the 4 + 2 = 6, 2 + 4 = 6, 6 − 4 = 2, and 6 − 2 = 4 cards.
- Say, **Let's make one more dot yellow** (to make the number sentence 3 + 3 = 6). Say, **What number sentence does this say?** Wait for a response.
- Hold up the 3 + 3 = 6 card and say, **See, if I say the yellow dots first, then the dots say the same sentence. 3 plus 3 equals 6. There is one minus sentence, too. What is it?** Wait for a response.

PARTNERS OF 7 ACTIVITIES

MATERIALS: Teacher Ten Frame Mat and 7 two-colored dots; Student Ten Frame Trays and 7 two-colored dots; white board magnetic easel; Number Sentence Cards: 6 + 1 = 7, 1 + 6 = 7, 7 − 6 = 1, 7 − 1 = 6, 5 + 2 = 7, 2 + 5 = 7, 7 − 5 = 2, 7 − 2 = 5, 4 + 3 = 7, 3 + 4 = 7, 7 − 4 = 3, 7 − 3 = 4

- Put the Teacher Ten Frame Mat on the white board magnetic easel. Add 7 red dots and say, **Here are 7 red dots. Let's see how many ways we can split 7 into two parts.**

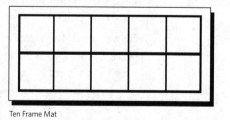
Ten Frame Mat

- Turn over the bottom dot farthest to the right.

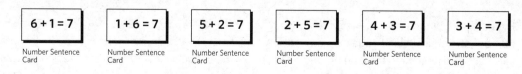

- Put down the plus cards: 6 + 1 = 7, 1 + 6 = 7, 5 + 2 = 7, 2 + 5 = 7, 4 + 3 = 7, and 3 + 4 = 7.

- Say, **Which card shows the number sentence 6 plus 1 equals 7?** Call on someone to choose the correct card. If he or she chooses the incorrect card, then hold the card up to the dots to compare. Allow him or her to choose another card.
- Point to the yellow dot and say, **I can put the yellow dot in the sentence first. Then I say 1 plus 6 equals 7.** Gesture to the dots in that order. **Which card says 1 plus 6 equals 7?** Call on someone to choose the correct card.

7 − 6 = 1	7 − 1 = 6	7 − 5 = 2	7 − 2 = 5	7 − 4 = 3	7 − 3 = 4
Number Sentence Card	Number Sentence Card	Number Sentence Card	Number Sentence Card	Number Sentence Card	Number Sentence Card

- Say, **Now let's find the number sentences that are minus.** Push the plus cards to the side and put down the minus cards: 7 − 6 = 1, 7 − 1 = 6, 7 − 5 = 2, 7 − 2 = 5, 7 − 4 = 3, and 7 − 3 = 4.
- Say, **First, I will cover the yellow dots.** Cover the 1 yellow dot.
- Say, **Which number sentence am I showing? 7 minus ____.** Gesture covering the 1 yellow dot. Call on someone to choose the correct card. Remove cards from the pile each time.
- Say, **Now I will cover the red dots.** Cover the 6 red dots.
- Say, **Which number sentence am I showing? 7 minus ____.** Call on someone to choose the correct card. Remove cards from the pile each time.
- Say, **Let's try another one! I am going to give you each a frame and 7 dots.** Pass out the Student Ten Frame Trays with 7 two-colored magnetic dots (5 red dots and 2 yellow dots). Keep 5 red dots on the top row, 1 red and 1 yellow on the bottom row. The dot farthest to the right on the bottom will be yellow.
- Change your dots to match the children's dots. Say, **What number sentence does this say? We already know that 5 plus 2 equals 7.** Gesture to the dots. **Let's show that on our fingers. 5 plus 2 equals 7.** Show on fingers. Make sure all the children show it correctly.
- Put down the plus cards. Say, **Which card shows the number sentence 5 plus 2 equals 7?** Call on someone to choose the correct Number Sentence Card. If he or she chooses the incorrect card, then hold the card up to the dots to compare. Allow him or her to choose another card.
- Say, **I can put the yellow dots in the first part of the sentence. Then I say 2 plus 5 equals 7.** Gesture first to the 2 yellow dots and then the 5 red dots. Say, **which card says 2 plus 5 equals 7?** Call on someone to choose the correct Number Sentence Card.
- Say, **Now let's find the number sentences that are minus.** Push the plus cards to the side and put down the minus cards.
- Say, **What number sentences go with these dots? There are 2.** Cover the 2 yellow dots.
- Say, **Which number sentence am I showing? 7 minus ____.** Gesture covering the 2 yellow dots. Call on someone to choose the correct card. Remove cards from the pile each time.
- Repeat covering the 5 red dots.
- Say, **Which number sentence am I showing? 7 minus ____.** Call on someone to choose the correct card. Remove cards from the pile each time.
- Say, **Let's try another one! We've done 1 yellow dot and 2 yellow dots. Let's try 3 yellow dots!** Say, **What number sentence does this say?** Repeat the previous protocol with 4 red and 3 yellow dots. Have the children copy you.

PARTNERS OF 8 ACTIVITIES

MATERIALS: Teacher Ten Frame Mat and 8 two-colored dots; Student Ten Frame Trays and 8 two-colored dots; white board magnetic easel; Number Sentence Cards: 7 + 1 = 8, 1 + 7 = 8, 8 − 7 =1, 8 − 1 = 7, 6 + 2 = 8, 2 + 6 = 8, 8 − 6 = 2, 8 − 2 = 6, 5 + 3 = 8, 3 + 5 = 8, 8 − 5 = 3, 8 − 3 = 5, 4 + 4 = 8, 8 − 4 = 4

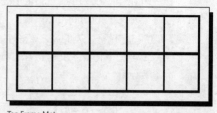
Ten Frame Mat

- Put the Teacher Ten Frame Mat on the white board magnetic easel. Add 8 red dots and say, **Here are 8 red dots. Let's see how many ways we can split 8 into two parts.**

- Turn over the bottom dot farthest to the right.

- Put down the plus cards: 7 + 1 = 8, 1 + 7 = 8, 6 + 2 = 8, 2 + 6 = 8, 5 + 3 = 8, 3 + 5 = 8, and 4 + 4 = 8.

- Say, **Which card shows the number sentence 7 plus 1 equals 8?** Call on someone to choose the correct card. If he or she chooses the incorrect card, then hold the card up to the dots to compare. Allow him or her to choose another card.

- Point to the yellow dot and say, **I can put the yellow dot in the sentence first. Then I say 1 plus 7 equals 8.** Gesture to the dots in that order. **Which card says 1 plus 7 equals 8?** Call on someone to choose the correct card.

- Say, **Now let's find the number sentences that are minus.** Push the plus cards to the side and put down the minus cards: 8 − 7 = 1, 8 − 1 = 7, 8 − 6 = 2, 8 − 2 = 6, 8 − 5 = 3, 8 − 3 = 5, and 8 − 4 = 4.

- Say, **First, I will cover the yellow dot.** Cover the 1 yellow dot.

- Say, **Which number sentence am I showing? 8 minus ____.** Gesture covering the 1 yellow dot. Call on someone to choose the correct card. Remove cards from the pile each time.

- Say, **Now I will cover the red dots.** Cover the 7 red dots.

- Say, **Which number sentence am I showing? 8 minus ____.** Gesture covering the 7 red dots. Call on someone to choose the correct card. Remove cards from the pile each time.

- Say, **Let's try another one! I am going to give you each a frame and 8 dots.** Pass out the Student Ten Frame Trays with 8 two-colored magnetic dots (6 red dots and 2 yellow dots). Keep 5 red dots on the top row, 1 red and 2 yellow dots on the bottom row. The 2 dots farthest to the right on the bottom will be yellow.

- Change your dots to match theirs. Say, **What number sentence does this say?**

- Put down the plus Number Sentence Cards for this activity.

- Say, **Which card shows the number sentence 6 plus 2 equals 8?** Call on someone to choose the correct card. If he or she chooses the incorrect card, then hold the card up to the dots to compare. Allow him or her to choose another card.

- Say, **I can put the yellow dots in the first part of the sentence. Then I say 2 plus 6 equals 8.** Gesture first to the yellow dots and then the red dots.

Number Sense Interventions Extension Partner Activities

- Say, **Which card says 2 plus 6 equals 8?** Call on someone to choose the correct card.
- Say, **Now let's find the number sentences that are minus.** Push the plus cards to the side and put down the minus cards.
- Say, **What number sentences go with these dots? There are 2.** Cover the 2 yellow dots.
- Say, **Which number sentence am I showing? 8 minus ____.** Gesture covering the 2 yellow dots. Call on someone to choose the correct Number Sentence Card. Remove cards from the pile each time.
- Repeat covering the 6 red dots.
- Say, **Which number sentence am I showing? 8 minus ____.** Gesture covering the 6 red dots. Call on someone to choose the correct Number Sentence Card. Remove cards from the pile each time.
- Say, **Let's try another one! We've done 1 yellow dot and 2 yellow dots. Let's try 3 yellow dots!** Turn the last red dot on the bottom row to the yellow side. Say, **What number sentence does this say?**
- Put down the plus Number Sentence Cards for this activity.
- Say, **Which card shows the number sentence 5 plus 3 equals 8.** Call on someone to choose the correct Number Sentence Card. If he or she chooses the incorrect card, then hold the card up to the dots to compare. Allow him or her to choose another card.
- Say, **I can put the yellow dots in the first part of the sentence. Then I say 3 plus 5 equals 8.** Gesture to the dots in that order.
- Say, **Which card says 3 plus 5 equals 8?** Call on someone to choose the correct Number Sentence Card.
- Say, **Now let's find the number sentences that are minus.** Push the plus cards to the side and put down the minus cards.
- Say, **What number sentences go with these dots? There are 2.** Cover the yellow dots.
- Say, **Which number sentence am I showing? 8 minus ____.** Call on someone to choose the correct Number Sentence Card. Remove cards from the pile each time.
- Repeat, covering the red dots.
- Say, **Which number sentence am I showing? 8 minus ____.** Call on someone to choose the correct Number Sentence Card.
- Turn over another yellow dot and say, **Let's turn over another dot.** Turn over the red dot farthest to the right. Say, **What number sentence does this say?** Wait for a response.
- Hold up the 4 + 4 = 8 card and say, **See, if I say the yellow dots first, the dots say the same sentence. 4 plus 4 equals 8. There is a similar minus sentence. What is it?** Wait for a response.

PARTNERS OF 9 ACTIVITIES

MATERIALS: Teacher Ten Frame Mat and 9 two-colored dots; Student Ten Frame Trays and 9 two-colored dots; white board magnetic easel; Number Sentence Cards: 8 + 1 = 9, 1 + 8 = 9, 9 − 8 = 1, 9 − 1 = 8, 7 + 2 = 9, 2 + 7 = 9, 9 − 7 = 2, 9 − 2 = 7, 6 + 3 = 9, 3 + 6 = 9, 9 − 6 = 3, 9 − 3 = 6, 5 + 4 = 9, 4 + 5 = 9, 9 − 5 = 4, 9 − 4 = 5

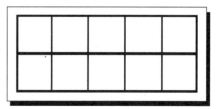
Ten Frame Mat

- Put the Teacher Ten Frame Mat on the white board magnetic easel. Add 9 red dots and say, **Here are 9 red dots. Let's see how many ways we can split 9 into two parts.**
- Turn over the bottom dot farthest to the right.

| 8 + 1 = 9 | 1 + 8 = 9 | 7 + 2 = 9 | 2 + 7 = 9 | 6 + 3 = 9 | 3 + 6 = 9 | 5 + 4 = 9 | 4 + 5 = 9 |

Number Sentence Card (×8)

- Put down the plus cards: 8 + 1 = 9, 1 + 8 = 9, 7 + 2 = 9, 2 + 7 = 9, 6 + 3 = 9, 3 + 6 = 9, 5 + 4 = 9, and 4 + 5 = 9.
- Say, **Which card shows the number sentence 8 plus 1 equals 9?** Call on someone to choose the correct card. If he or she chooses the incorrect card, then hold the card up to the dots to compare. Allow him or her to choose another card.
- Point to the yellow dot and say, **I can put the yellow dot in the first part of the sentence. Then I say 1 plus 8 equals 9.** Gesture to the dots in that order. **Which card says 1 plus 8 equals 9?** Call on someone to choose the correct Number Sentence Card.

9 − 8 = 1	9 − 1 = 8	9 − 7 = 2	9 − 2 = 7	9 − 6 = 3	9 − 3 = 6	9 − 5 = 4	9 − 4 = 5
Number Sentence Card	Number Sentence Card	Number Sentence Card	Number Sentence Card	Number Sentence Card	Number Sentence Card	Number Sentence Card	Number Sentence Card

- Say, **Now let's find the number sentences that are minus.** Push the plus cards to the side and put down the minus cards: 9 − 8 = 1, 9 − 1 = 8, 9 − 7 = 2, 9 − 2 = 7, 9 − 6 = 3, 9 − 3 = 6, 9 − 5 = 4, 9 − 4 = 5.
- Say, **First, I will cover the yellow dots.** Cover the 1 yellow dot.
- Say, **Which number sentence am I showing? 9 minus ____.** Gesture covering the 1 yellow dot. Call on someone to choose the correct Number Sentence Card. Remove cards from the pile each time.
- Say, **Now I will cover the red dots.** Cover the 8 red dots.
- Say, **Which number sentence am I showing? 9 minus ____.** Gesture covering the 8 red dots. Call on someone to choose the correct Number Sentence Card. Remove cards from the pile each time.
- Say, **Let's try another one! I am going to give you each a frame and 9 dots.** Pass out the Student Ten Frame Trays with 9 two-colored magnetic dots (7 red dots and 2 yellow dots). Keep 5 red dots on the top row, 2 red and 2 yellow on the bottom row. The 2 dots farthest to the right on the bottom will be yellow.
- Change your dots to match theirs. Say, **What number sentence does this say?**
- Put down the plus Number Sentence Cards for this activity.
- Say, **Which card shows the number sentence 7 plus 2 equals 9?** Call on someone to choose the correct Number Sentence Card. If he or she chooses the incorrect card, then hold the card up to the dots to compare. Allow him or her to choose another card.
- Say, **I can put the yellow dots in the first part of the sentence. Then I say 2 plus 7 equals 9.** Gesture first to the yellow dots and then the red dots.
- Say, **Which card says 2 plus 7 equals 9?** Call on someone to choose the correct Number Sentence Card.
- Say, **Now let's find the number sentences that are minus.** Push the plus cards to the side and put down the minus number sentence cards for this activity.
- Say, **What number sentences go with these dots? There are 2.** Cover the 2 yellow dots.
- Say, **Which number sentence am I showing? 9 minus ____.** Call on someone to choose the correct Number Sentence Card. Remove cards from the pile each time.
- Repeat covering the 7 red dots.
- Say, **Which number sentence am I showing? 9 minus ____.** Gesture covering the 7 red dots. Call on someone to choose the correct Number Sentence Card. Remove cards from the pile each time.
- Say, **Let's try another one! We've done 1 yellow dot and 2 yellow dots. Let's try 3 yellow dots!** Turn the next red dot on the bottom row to the yellow side. Say, **What number sentence does this say?**
- Put down the plus Number Sentence Cards for this activity.
- Say, **Which card shows the number sentence 6 plus 3 equals 9?** Call on someone to choose the correct Number Sentence Card. If he or she chooses the incorrect card, then hold the card up to the dots to compare. Allow him or her to choose another card.

Number Sense Interventions Extension Partner Activities

- Say, **I can put the yellow dots in the first part of the sentence. Then I say 3 plus 6 equals 9.** Gesture to the dots in that order.
- Say, **Which card says 3 plus 6 equals 9?** Call on someone to choose the correct Number Sentence Card.
- Say, **Now let's find the number sentences that are minus.** Push the plus cards to the side and put down the minus cards.
- Say, **What number sentences go with these dots? There are 2.** Cover the 3 yellow dots.
- Say, **Which number sentence am I showing? 9 minus ____.** Call on someone to choose the correct Number Sentence Card. Remove cards from the pile each time.
- Repeat, covering the 6 red dots.
- Say, **Which number sentence am I showing? 9 minus ____.** Call on someone to choose the correct Number Sentence Card.
- Turn over the bottom 4 dots. **We already know that 5 plus 4 equals 9.** Gesture to the dots. **Let's show that on our fingers. 5 plus 4 equals 9.** Show on fingers. Make sure all the children show it correctly.
- Put down the plus Number Sentence Cards for this activity. Say, **Which card shows the number sentence 5 plus 4 equals 9?** Call on someone to choose the correct card. If he or she chooses the incorrect card, then hold the card up to the dots to compare. Allow him or her to choose another card.
- Say, **I can put the yellow dots in the first part of the sentence. Then I say 4 plus 5 equals 9.** Gesture to the dots in that order.
- Say, **Which card says 4 plus 5 equals 9?** Call on someone to choose the correct Number Sentence Card.
- Say, **Now let's find the number sentences that are minus.** Push the plus number sentence cards to the side and put down the minus number sentence cards for this activity.
- Say, **What number sentences go with these dots? There are 2.** Cover the 4 yellow dots.
- Say, **Which number sentence am I showing? 9 minus ____.** Gesture covering the 4 yellow dots. Call on someone to choose the correct Number Sentence Card. Remove used cards from the pile each time.
- Repeat, covering the 5 red dots.

PARTNERS OF 10 ACTIVITIES

MATERIALS: Teacher Ten Frame Mat and 10 two-colored dots; Student Ten Frame Trays and 10 two-colored dots; white board magnetic easel; Number Sentence Cards: 9 + 1 = 10, 1 + 9 = 10, 10 − 9 = 1, 10 − 1 = 9, 8 + 2 = 10, 2 + 8 = 10, 10 − 8 = 2, 10 − 2 = 8, 7 + 3 = 10, 3 + 7 = 10, 10 − 7 = 3, 10 − 3 = 7, 6 + 4 = 10, 4 + 6 = 10, 10 − 6 = 4, 10 − 4 = 6, 5 + 5 = 10, 10 − 5 = 5

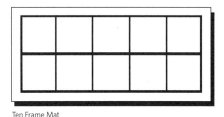
Ten Frame Mat

- Put the Teacher Ten Frame Mat on the white board magnetic easel. Add 10 red dots and say, **Here are 10 red dots. Let's see how many ways we can split 10 into two parts.**
- Turn over the bottom 5 dots. **We already know that 5 plus 5 equals 10.** Gesture to the dots. **Let's show that on our fingers. 5 plus 5 equals 10.** Show the number on your fingers. Make sure all the children show it correctly.

9 + 1 = 10	1 + 9 = 10	8 + 2 = 10	2 + 8 = 10	7 + 3 = 10
Number Sentence Card	Number Sentence Card	Number Sentence Card	Number Sentence Card	Number Sentence Card

3 + 7 = 10	6 + 4 = 10	4 + 6 = 10	5 + 5 = 10
Number Sentence Card	Number Sentence Card	Number Sentence Card	Number Sentence Card

- Put down the plus cards: 9 + 1 = 10, 1 + 9 = 10, 8 + 2 = 10, 2 + 8 = 10, 7 + 3 = 10, 3 + 7 = 10, 6 + 4 = 10, 4 + 6 = 10, and 5 + 5 = 10.
- Say, **Which card shows the number sentence 5 plus 5 equals 10?** Call on someone to choose the correct Number Sentence Card. If he or she chooses the incorrect card, then hold the card up to the dots to compare. Allow him or her to choose another card.
- Hold up the 5 + 5 = 10 card and say, **See, if I say the yellow dots first, the dots say the same sentence. 5 plus 5 equals 10. There is only one minus sentence, too. What is it?** Wait for a response.
- Change your dots to match the students' dots.
- Say, **Today you are each going to make partners for 10. You know how to do it now.** Pass out Student Ten Frame Trays with 10 magnetic dots (all red dots).
- Say, **Make some dots yellow and some red. Keep the same colors together. Do not mix them up.**
- When they are finished, ask, **Let's see what you have made!**
- Have each child show his or her tray. Say, **Are any trays the same?**
- Find any matching trays.
- Follow the previous protocol for finding number sentences that match each tray. If they make partners that were already made, then simply say, **That is good, but we already made those. Let's try another one.**
- Repeat until all Number Sentence Cards have been used.

CHAPTER 3

Photocopiable Activity Sheets

Lesson 1 Activity Sheet 146
Lesson 2 Activity Sheet 148
Lesson 3 Activity Sheet 150
Lesson 4 Activity Sheet 152
Lesson 5 Activity Sheet 154
Lesson 6 Activity Sheet 156
Lesson 7 Activity Sheet 158
Lesson 8 Activity Sheet 160
Lesson 9 Activity Sheet 162
Lesson 10 Activity Sheet 164
Lesson 11 Activity Sheet 166
Lesson 12 Activity Sheet 168
Lesson 13 Activity Sheet 170
Lesson 14 Activity Sheet 172
Lesson 15 Activity Sheet 174
Lesson 16 Activity Sheet 176
Lesson 17 Activity Sheet 177
Lesson 18 Activity Sheet 178
Lessons 19-24 Activity Sheet 179

LESSON 1 Activity Sheet

(page 1 of 2)

Name: _____ School: _____

146 Number Sense Interventions by Nancy C. Jordan and Nancy Dyson
Copyright © 2014 by Paul H. Brookes Publishing Co., Inc. All rights reserved.

LESSON 1 Activity Sheet

(page 2 of 2)

2 − 2 = 0	2 − 1 = 1
1 + 1 = 2	1 − 1 = 0

LESSON 2 Activity Sheet   *(page 1 of 2)*

Name: _____   School: _____

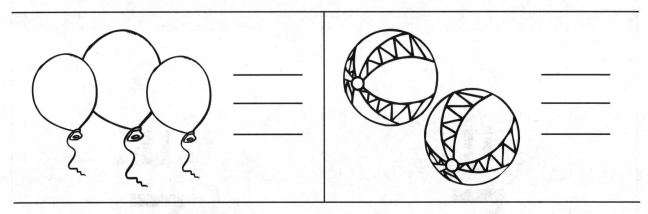

_____
_____

_____
_____

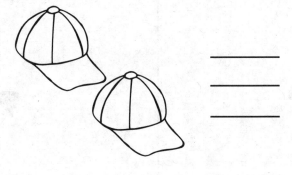

_____
_____

_____

LESSON 2 Activity Sheet *(page 2 of 2)*

$1 + 2 = \_$	$3 - 1 = \_$
$1 + 1 = \_$	$3 - 2 = \_$
$2 + 1 = \_$	$3 - 3 = \_$

Number Sense Interventions by Nancy C. Jordan and Nancy Dyson
Copyright © 2014 by Paul H. Brookes Publishing Co., Inc. All rights reserved.

LESSON 3 Activity Sheet

(page 1 of 2)

Name:_____ School:_____

3 3 3 3   4 4 4 4

3   4

LESSON 3 Activity Sheet

(page 2 of 2)

3 + 1 = _	4 − 3 = _
1 + 3 = _	4 − 1 = _
1 + 1 = _	3 − 2 = _
2 + 1 = _	2 − 1 = _

LESSON 4 Activity Sheet

(page 1 of 2)

Name: _____ School: _____

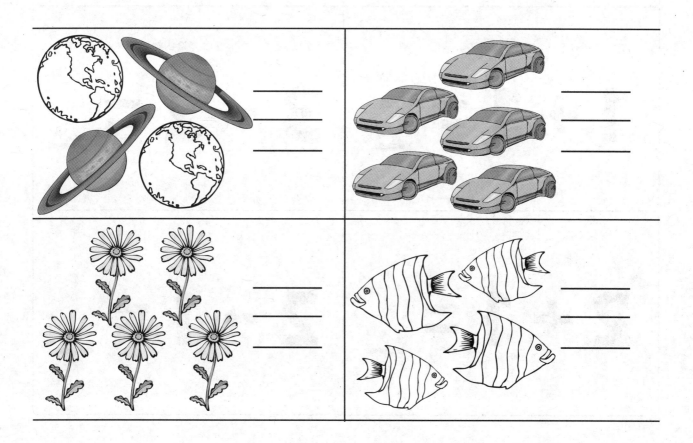

LESSON 4 Activity Sheet

2 + 2 = _	4 − 2 = _
3 + 1 = _	2 − 2 = _
1 + 2 = _	3 − 2 = _
1 + 3 = _	2 − 1 = _

LESSON 5 Activity Sheet

(page 1 of 2)

Name: _____ School: _____

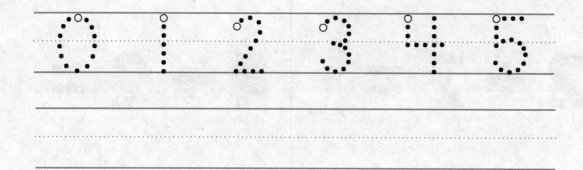

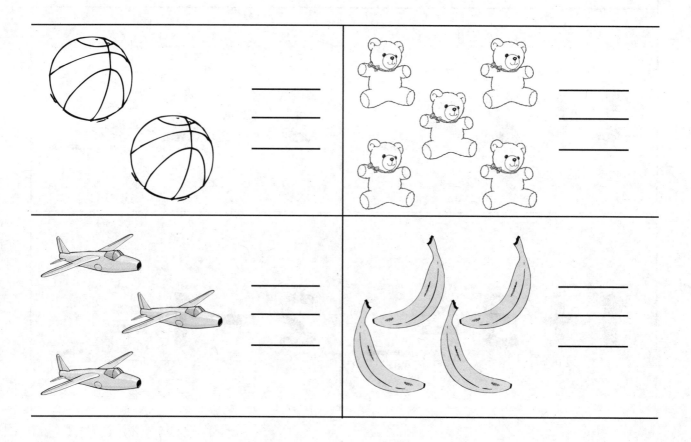

LESSON 5 Activity Sheet (page 2 of 2)

2 + 3 = _	5 − 3 = _
3 + 2 = _	5 − 2 = _
1 + 4 = _	5 − 4 = _
4 + 1 = _	5 − 1 = _

LESSON 6 Activity Sheet

(page 1 of 2)

Name: _____ School: _____

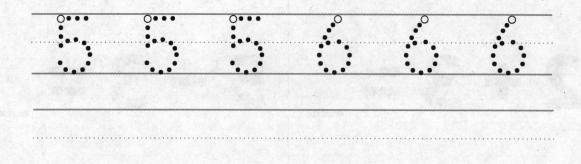

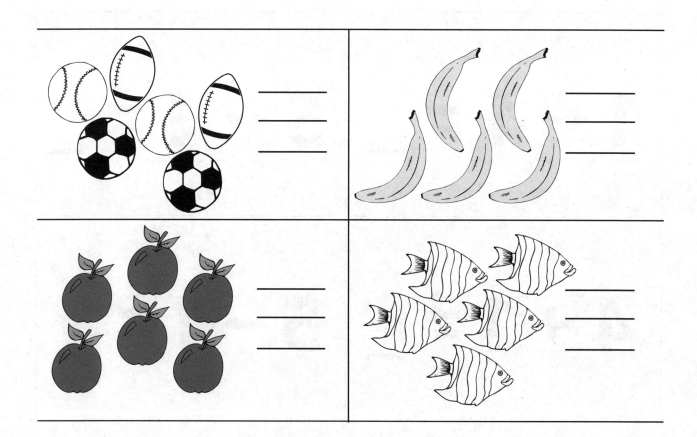

LESSON 6 Activity Sheet

2 + _ = 5	5 + 0 = _
1 + 4 = _	3 + _ = 5
2 + 2 = _	3 + 1 = _
_ + 1 = 5	0 + 5 = _

LESSON 7 Activity Sheet

Name: _____ School: _____

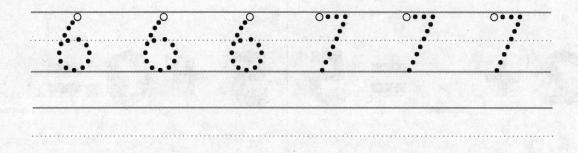

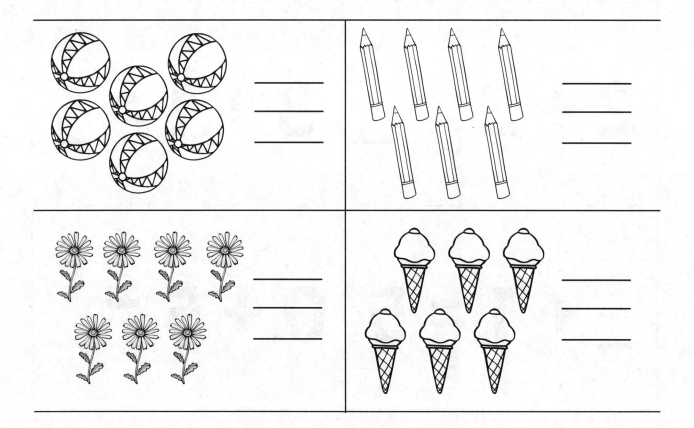

LESSON 7 Activity Sheet

5 - 0 = _	5 - _ = 4
5 - _ = 3	4 - 1 = _
5 - 4 = _	5 - 5 = _
3 - _ = 2	5 - 2 = _

LESSON 8 Activity Sheet

Name: _____ School: _____

7 7 7 8 8 8

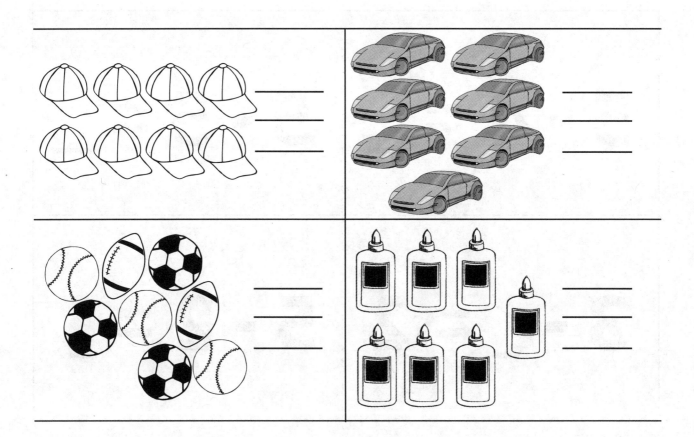

LESSON 8 Activity Sheet

(page 2 of 2)

$2 + 3 = \_$	$4 + 1 = \_$
$5 - \_ = 2$	$5 - 2 = \_$
$1 + 4 = \_$	$3 + \_ = 5$
$5 - 3 = \_$	$5 - \_ = 4$

LESSON 9 Activity Sheet

Name: _____ School: _____

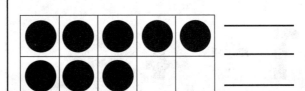

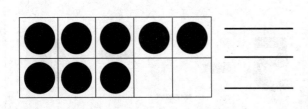

LESSON 9 Activity Sheet

2 + 2 = _	1 + 4 = _
3 + _ = 4	2 + 3 = _
3 + 2 = _	2 + 1 = _
5 + 0 = _	1 + _ = 2

LESSON 10 Activity Sheet

(page 1 of 2)

Name: _____   School: _____

9 9 9 10 10 10

LESSON 10 Activity Sheet

$5 - 5 = \_$	$5 - 1 = \_$
$5 - 3 = \_$	$5 - 2 = \_$
$5 - 4 = \_$	$5 - \_ = 1$
$3 - \_ = 2$	$4 - \_ = 2$

LESSON 11 Activity Sheet

(page 1 of 2)

Name: _____ School: _____

| 1 | 2 | 3 | 4 | | 6 | 7 | 8 | | 10 |

| 1 | | 3 | 4 | 5 | 6 | | 8 | 9 | 10 |

| 1 | 2 | | | 5 | 6 | 7 | 8 | 9 | 10 |

| | 2 | 3 | 4 | 5 | 6 | 7 | 8 | 9 | |

LESSON 11 Activity Sheet

(page 2 of 2)

5 + _ = 5	1 + _ = 2
2 + _ = 3	4 + _ = 5
1 + 3 = _	3 + 2 = _
5 - _ = 4	4 - _ = 2
1 - _ = 0	2 - 1 = _
3 - 2 = _	5 - 4 = _

LESSON 12 Activity Sheet

(page 1 of 2)

Name: _____ School: _____

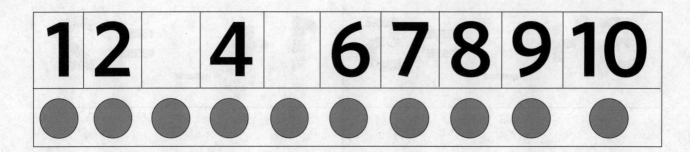

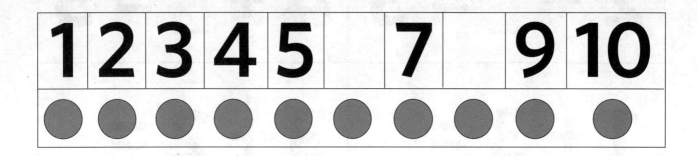

168 *Number Sense Interventions* by Nancy C. Jordan and Nancy Dyson
Copyright © 2014 by Paul H. Brookes Publishing Co., Inc. All rights reserved.

LESSON 12 Activity Sheet

(page 2 of 2)

$5 - \_ = 3$	$2 + 0 = \_$
$5 - 3 = \_$	$2 - \_ = 1$
$2 + 0 = \_$	$3 - 1 = \_$
$5 - \_ = 2$	$4 - 2 = \_$
$3 + 2 = \_$	$3 - \_ = 1$
$2 - \_ = 2$	$4 + 1 = \_$

LESSON 13 Activity Sheet   (page 1 of 2)

Name: _____   School: _____

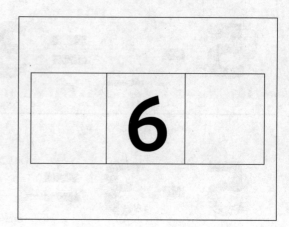

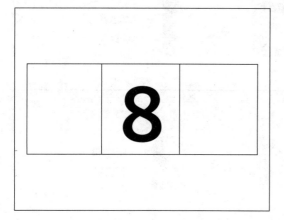

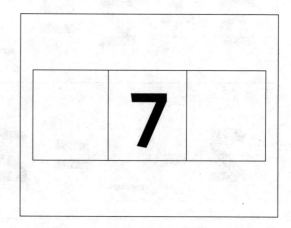

LESSON 13 Activity Sheet (page 2 of 2)

4 + 1 = _	5 − 4 = _
3 + 2 = _	2 − 0 = _
2 + 0 = _	5 − 0 = _
3 + 1 = _	4 − 3 = _

LESSON 14 Activity Sheet

Name: _____ School: _____

4 + 1	4 + 2	3 - 1
3 - 2	3 + 1	2 + 2
4 - 1	4 - 2	5 + 1
3 + 2	5 - 1	5 - 2

LESSON 14 Activity Sheet

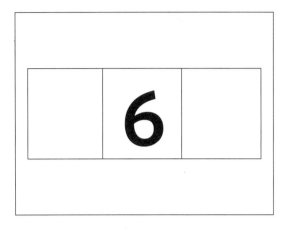

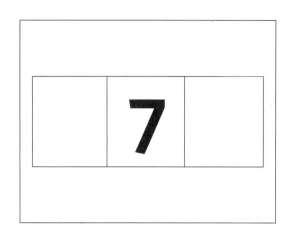

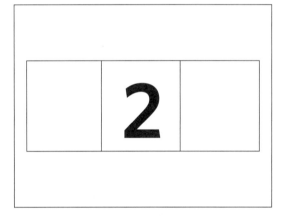

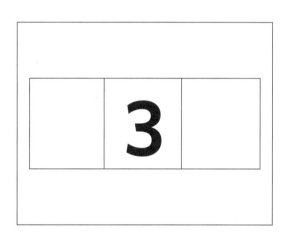

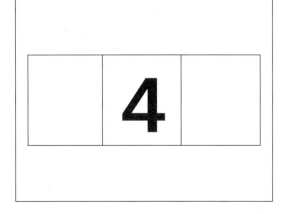

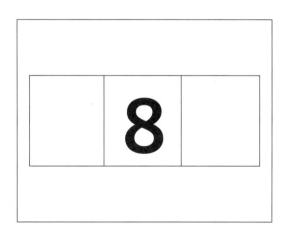

LESSON 15 Activity Sheet

Name:_____ School: _____

$3 \\ +1$	$5 \\ +1$	$2 \\ -1$
$6 \\ -1$	$7 \\ +1$	$4 \\ +1$
$4 \\ -1$	$8 \\ -1$	$9 \\ +1$
$6 \\ +1$	$5 \\ -1$	$3 \\ -1$

LESSON 15 Activity Sheet

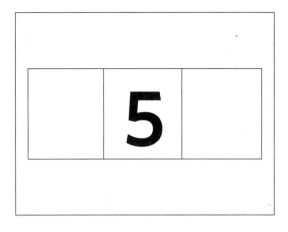
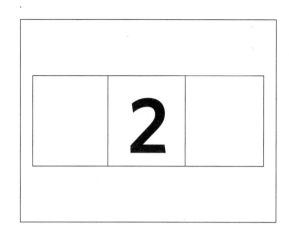
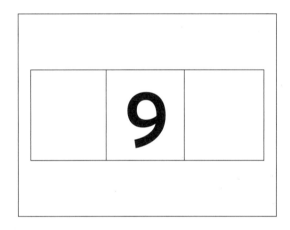
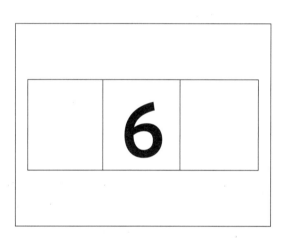
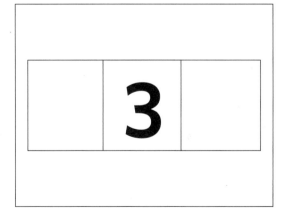
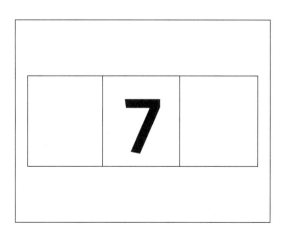

LESSON 16 Activity Sheet

(page 1 of 1)

Name: _____ School: _____

176 *Number Sense Interventions* by Nancy C. Jordan and Nancy Dyson
Copyright © 2014 by Paul H. Brookes Publishing Co., Inc. All rights reserved.

LESSON 17 Activity Sheet

(page 1 of 1)

Name: _____ School: _____

Number Sense Interventions by Nancy C. Jordan and Nancy Dyson
Copyright © 2014 by Paul H. Brookes Publishing Co., Inc. All rights reserved.

LESSON 18 Activity Sheet

(page 1 of 1)

Name:_____ School:_____

178

Number Sense Interventions by Nancy C. Jordan and Nancy Dyson
Copyright © 2014 by Paul H. Brookes Publishing Co., Inc. All rights reserved.

LESSONS 19–24 Activity Sheet

(page 1 of 1)

Name:_____ School:_____

CHAPTER 4

Photocopiable Materials

Cardinality Chart	182
Subitizing Circle Cards	183
Dot Chart for 2	190
Dot Chart for 3	191
Dot Chart for 4	192
Number Sentence Cards	193
Partner Dot Cards	210
Hundreds Chart	211
Five Frames Master	212
Ten Frames Master	213
Decade Cards	214
Unit Cards	220
Bigger/Smaller Cards	223
Teacher Number List	224
Ten Frame Flash Cards	225
Student Number List	226
Activity 18 Hundreds Chart	227
Activity 19 Hundreds Chart	228
Activity 20 Hundreds Chart	229
Activity 21 Hundreds Chart	230
Activity 22 Hundreds Chart	231
Activity 23 Hundreds Chart	232
Activity 24 Hundreds Chart	233

Cardinality Chart

(page 1 of 1)

| 1 | 2 | 3 | 4 | 5 | 6 | 7 | 8 | 9 | 10 |

182 Number Sense Interventions by Nancy C. Jordan and Nancy Dyson
Copyright © 2014 by Paul H. Brookes Publishing Co., Inc. All rights reserved.

Subitizing Circle Cards

(page 1 of 7)

1	2
SUBITIZING CIRCLE CARDS	SUBITIZING CIRCLE CARDS

3	4
SUBITIZING CIRCLE CARDS	SUBITIZING CIRCLE CARDS

Subitizing Circle Cards

5

SUBITIZING CIRCLE CARDS

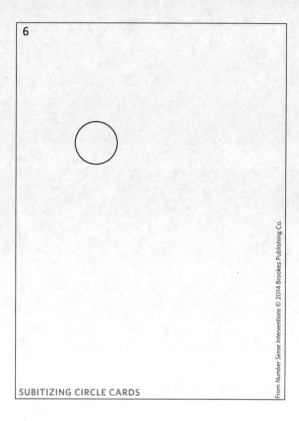

6

SUBITIZING CIRCLE CARDS

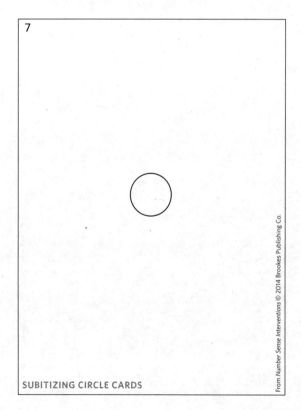

7

SUBITIZING CIRCLE CARDS

8

SUBITIZING CIRCLE CARDS

Subitizing Circle Cards

(page 3 of 7)

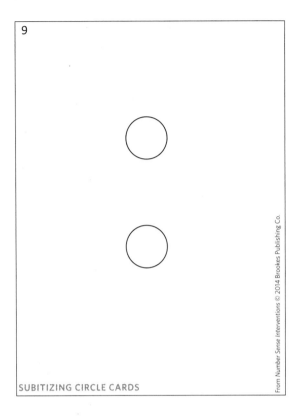

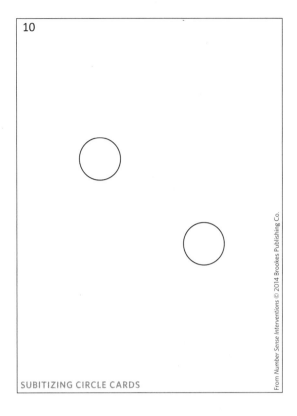

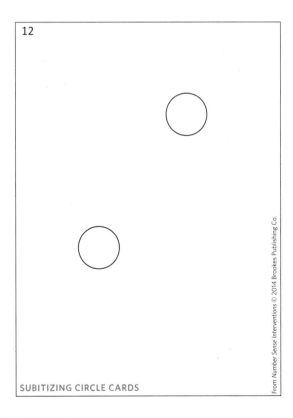

Number Sense Interventions by Nancy C. Jordan and Nancy Dyson
Copyright © 2014 by Paul H. Brookes Publishing Co., Inc. All rights reserved.

Subitizing Circle Cards

(page 4 of 7)

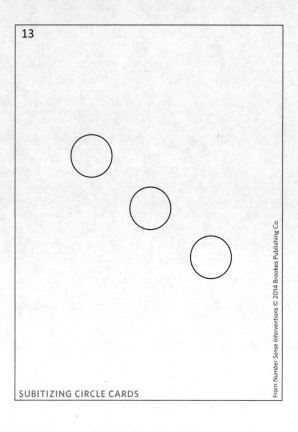

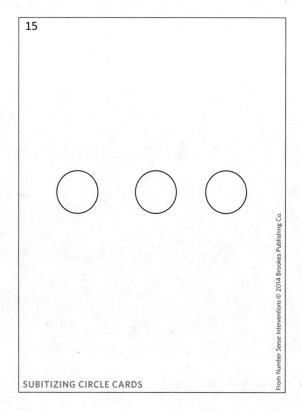

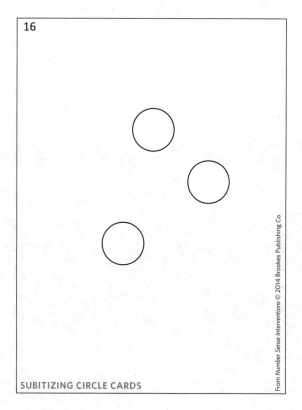

Subitizing Circle Cards

(page 5 of 7)

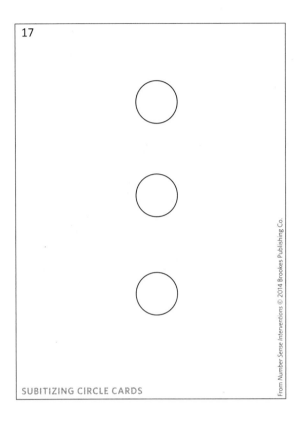
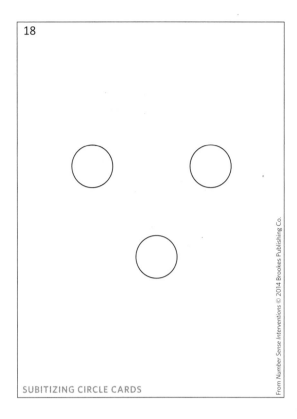
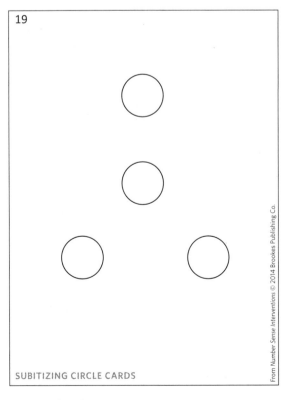
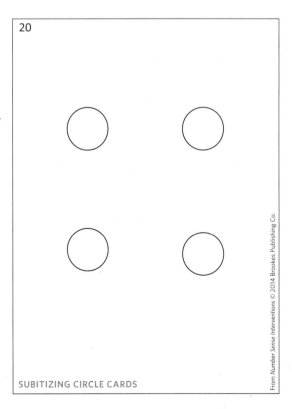

Number Sense Interventions by Nancy C. Jordan and Nancy Dyson
Copyright © 2014 by Paul H. Brookes Publishing Co., Inc. All rights reserved.

187

Subitizing Circle Cards

(page 6 of 7)

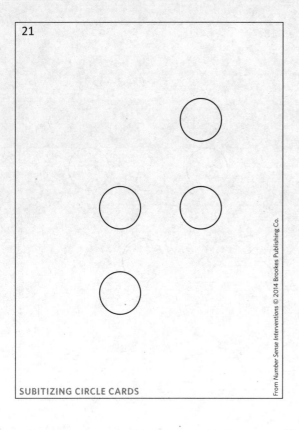
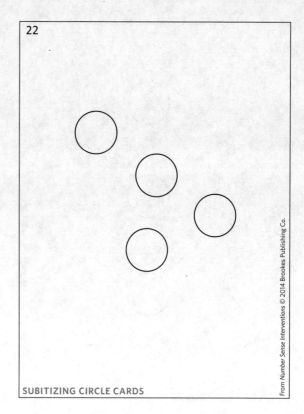
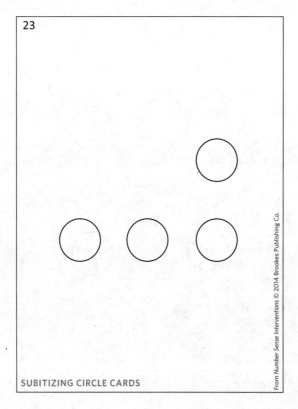
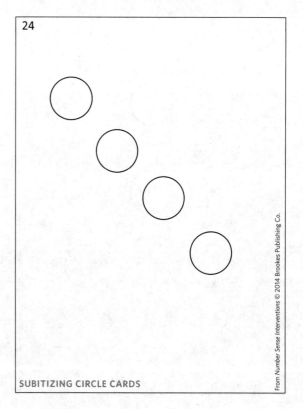

Subitizing Circle Cards

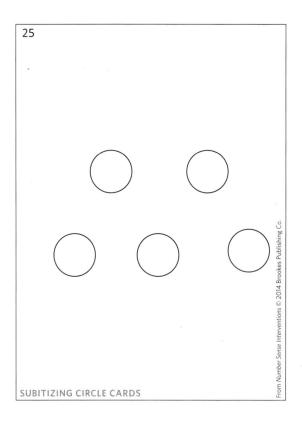

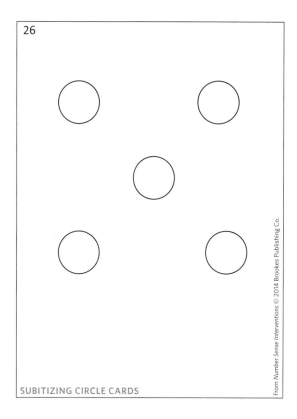

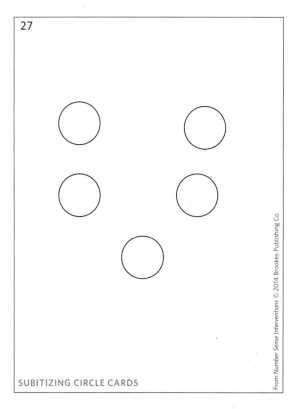

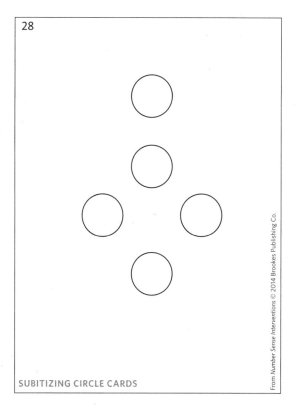

Dot Chart for 2

Dot Chart for 3

Dot Chart for 4

Number Sentence Cards

(page 1 of 17)

1 + 2 = 3	1 + 3 = 4	2 + 2 = 4
NUMBER SENTENCE CARDS	NUMBER SENTENCE CARDS	NUMBER SENTENCE CARDS
1 + 1 = 2	2 + 1 = 3	3 + 1 = 4
NUMBER SENTENCE CARDS	NUMBER SENTENCE CARDS	NUMBER SENTENCE CARDS

Number Sense Interventions by Nancy C. Jordan and Nancy Dyson
Copyright © 2014 by Paul H. Brookes Publishing Co., Inc. All rights reserved.

Number Sentence Cards

$4 + 1 = 5$	$3 + 2 = 5$	$4 - 1 = 3$
NUMBER SENTENCE CARDS	NUMBER SENTENCE CARDS	NUMBER SENTENCE CARDS
$1 + 4 = 5$	$2 + 3 = 5$	$5 - 1 = 4$
NUMBER SENTENCE CARDS	NUMBER SENTENCE CARDS	NUMBER SENTENCE CARDS

Number Sentence Cards

(page 3 of 17)

2 − 1 = 1	4 − 2 = 2	5 − 3 = 2
NUMBER SENTENCE CARDS	NUMBER SENTENCE CARDS	NUMBER SENTENCE CARDS
3 − 1 = 2	5 − 2 = 3	3 − 2 = 1
NUMBER SENTENCE CARDS	NUMBER SENTENCE CARDS	NUMBER SENTENCE CARDS

Number Sentence Cards

5 − 4 = 1	1 + 5 = 6	2 + 4 = 6
4 − 3 = 1	3 + 3 = 6	5 + 1 = 6

Number Sentence Cards

(page 5 of 17)

$6 - 1 = 5$	$6 - 3 = 3$	$6 - 5 = 1$
NUMBER SENTENCE CARDS	NUMBER SENTENCE CARDS	NUMBER SENTENCE CARDS
$4 + 2 = 6$	$6 - 2 = 4$	$6 - 4 = 2$
NUMBER SENTENCE CARDS	NUMBER SENTENCE CARDS	NUMBER SENTENCE CARDS

Number Sense Interventions by Nancy C. Jordan and Nancy Dyson
Copyright © 2014 by Paul H. Brookes Publishing Co., Inc. All rights reserved.

Number Sentence Cards

6 + 1 = 7	5 + 2 = 7	4 + 3 = 7
NUMBER SENTENCE CARDS	NUMBER SENTENCE CARDS	NUMBER SENTENCE CARDS
1 + 6 = 7	2 + 5 = 7	3 + 4 = 7
NUMBER SENTENCE CARDS	NUMBER SENTENCE CARDS	NUMBER SENTENCE CARDS

Number Sentence Cards

(page 7 of 17)

$7 - 5 = 2$	$7 - 3 = 4$	$7 - 1 = 6$
NUMBER SENTENCE CARDS	NUMBER SENTENCE CARDS	NUMBER SENTENCE CARDS
$7 - 6 = 1$	$7 - 4 = 3$	$7 - 2 = 5$
NUMBER SENTENCE CARDS	NUMBER SENTENCE CARDS	NUMBER SENTENCE CARDS

Number Sense Interventions by Nancy C. Jordan and Nancy Dyson
Copyright © 2014 by Paul H. Brookes Publishing Co., Inc. All rights reserved.

Number Sentence Cards

(page 8 of 17)

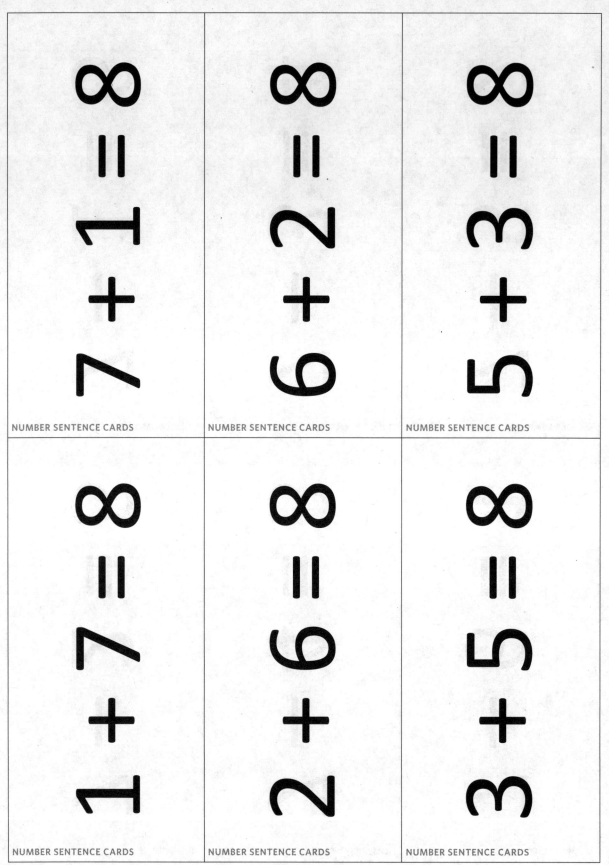

Number Sentence Cards

8 − 7 = 1	8 − 5 = 3	8 − 3 = 5
NUMBER SENTENCE CARDS	NUMBER SENTENCE CARDS	NUMBER SENTENCE CARDS
4 + 4 = 8	8 − 6 = 2	8 − 4 = 4
NUMBER SENTENCE CARDS	NUMBER SENTENCE CARDS	NUMBER SENTENCE CARDS

Number Sense Interventions by Nancy C. Jordan and Nancy Dyson
Copyright © 2014 by Paul H. Brookes Publishing Co., Inc. All rights reserved.

Number Sentence Cards

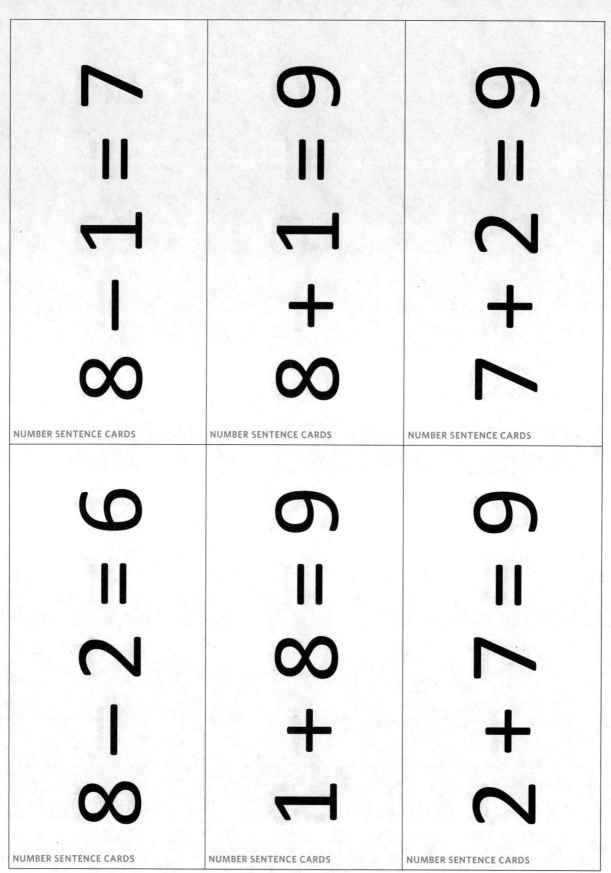

Number Sentence Cards

(page 11 of 17)

$6 + 3 = 9$	$5 + 4 = 9$	$9 - 7 = 2$
$3 + 6 = 9$	$4 + 5 = 9$	$9 - 8 = 1$

Number Sentence Cards

(page 12 of 17)

$9-5=4$	$9-3=6$	$9-1=8$
$9-6=3$	$9-4=5$	$9-2=7$

Number Sentence Cards

9 + 1 = 10	8 + 2 = 10	7 + 3 = 10
1 + 9 = 10	2 + 8 = 10	3 + 7 = 10

Number Sentence Cards

6 + 4 = 10	10 − 9 = 1	10 − 7 = 3
NUMBER SENTENCE CARDS	NUMBER SENTENCE CARDS	NUMBER SENTENCE CARDS
4 + 6 = 10	5 + 5 = 10	10 − 8 = 2
NUMBER SENTENCE CARDS	NUMBER SENTENCE CARDS	NUMBER SENTENCE CARDS

Number Sentence Cards

(page 15 of 17)

$10 - 5 = 5$	$10 - 3 = 7$	$10 - 1 = 9$
NUMBER SENTENCE CARDS	NUMBER SENTENCE CARDS	NUMBER SENTENCE CARDS
$10 - 6 = 4$	$10 - 4 = 6$	$10 - 2 = 8$
NUMBER SENTENCE CARDS	NUMBER SENTENCE CARDS	NUMBER SENTENCE CARDS

Number Sense Interventions by Nancy C. Jordan and Nancy Dyson
Copyright © 2014 by Paul H. Brookes Publishing Co., Inc. All rights reserved.

Number Sentence Cards

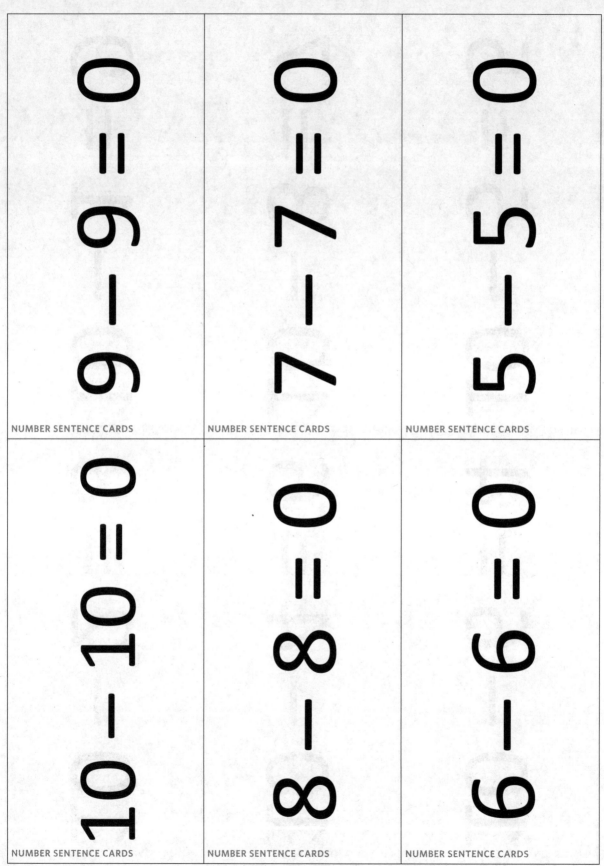

Number Sentence Cards

$3 - 3 = 0$

NUMBER SENTENCE CARDS

$1 - 1 = 0$

NUMBER SENTENCE CARDS

$4 - 4 = 0$

NUMBER SENTENCE CARDS

$2 - 2 = 0$

NUMBER SENTENCE CARDS

Partner Dot Cards

(page 1 of 1)

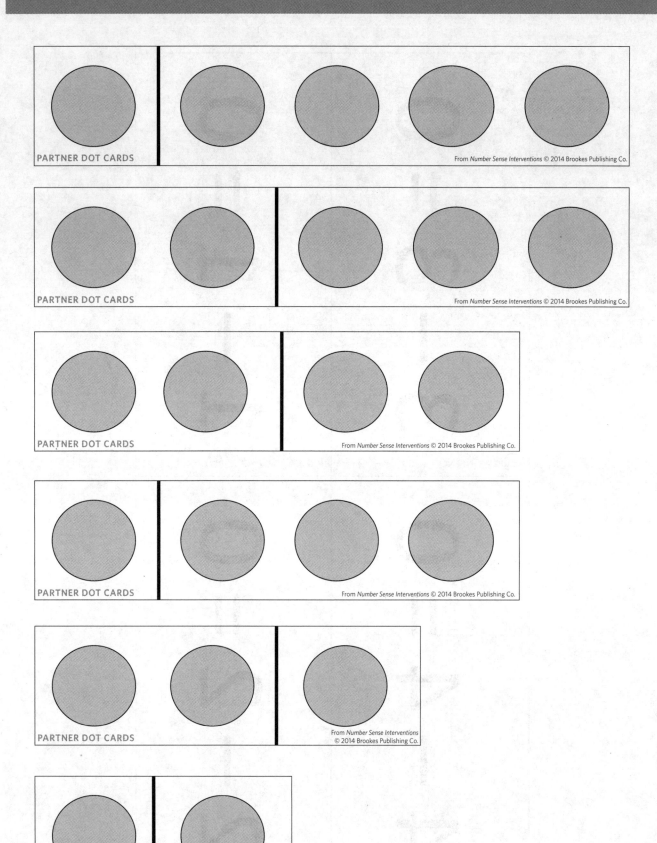

Hundreds Chart

1	2	3	4	5	6	7	8	9	10
11	12	13	14	15	16	17	18	19	20
21	22	23	24	25	26	27	28	29	30
31	32	33	34	35	36	37	38	39	40
41	42	43	44	45	46	47	48	49	50
51	52	53	54	55	56	57	58	59	60
61	62	63	64	65	66	67	68	69	70
71	72	73	74	75	76	77	78	79	80
81	82	83	84	85	86	87	88	89	90
91	92	93	94	95	96	97	98	99	**100**

Five Frames Master

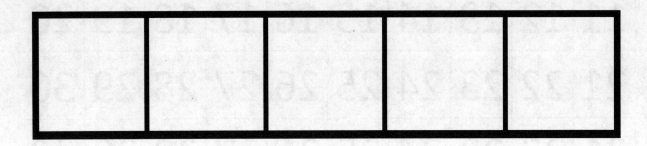

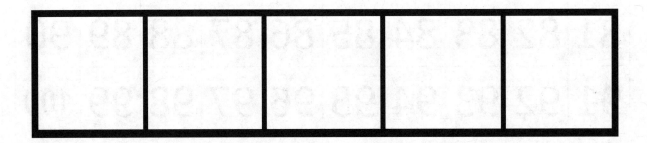

Ten Frames Master

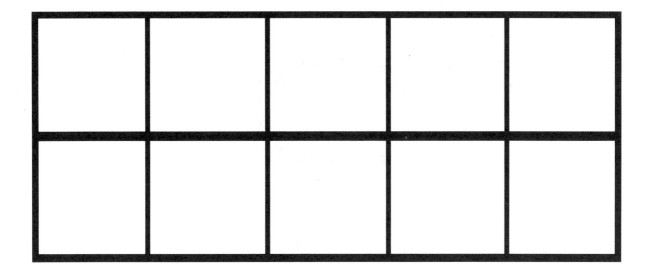

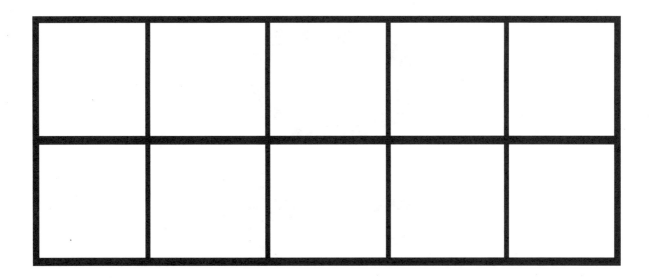

Decade Cards (page 1 of 6)

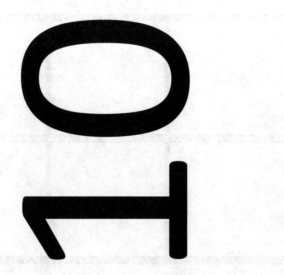

Decade Cards *(page 2 of 6)*

Decade Cards

(page 3 of 6)

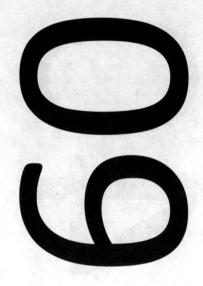

Decade Cards

(page 4 of 6)

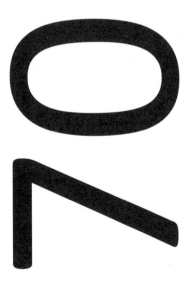

Decade Cards

90

Decade Cards

(page 6 of 6)

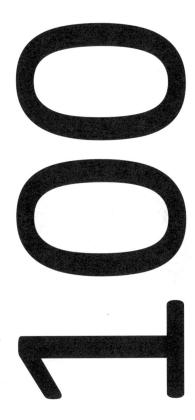

Unit Cards

(page 1 of 3)

Unit Cards (page 2 of 3)

6

5

4

Number Sense Interventions by Nancy C. Jordan and Nancy Dyson
Copyright © 2014 by Paul H. Brookes Publishing Co., Inc. All rights reserved.

221

Unit Cards (page 3 of 3)

Bigger/Smaller Cards

(page 1 of 1)

8 6	2 5
3 4	7 9
4 9	10 6
5 3	2 1
1 7	3 10

Number Sense Interventions by Nancy C. Jordan and Nancy Dyson
Copyright © 2014 by Paul H. Brookes Publishing Co., Inc. All rights reserved.

Teacher Number List

(page 1 of 1)

(ENLARGE AND PRINT ON 11 X 17 PAPER)

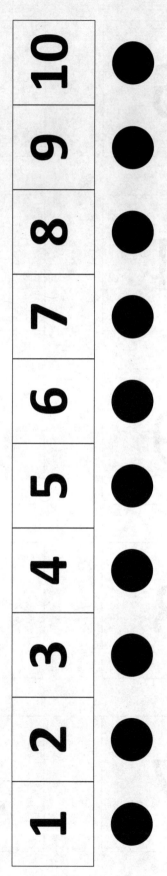

224 *Number Sense Interventions* by Nancy C. Jordan and Nancy Dyson
Copyright © 2014 by Paul H. Brookes Publishing Co., Inc. All rights reserved.

Ten Frame Flash Cards

(page 1 of 1)

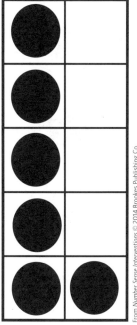
Ten Frame Flash Cards

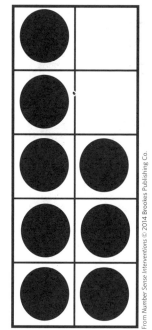
Ten Frame Flash Cards

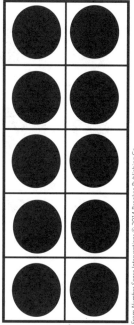
Ten Frame Flash Cards

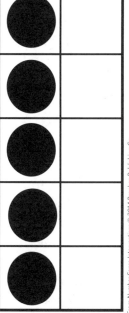
Ten Frame Flash Cards

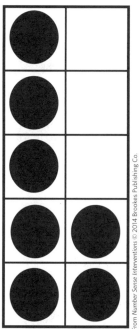
Ten Frame Flash Cards

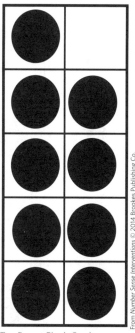
Ten Frame Flash Cards

Number Sense Interventions by Nancy C. Jordan and Nancy Dyson
Copyright © 2014 by Paul H. Brookes Publishing Co., Inc. All rights reserved.

Student Number List

(page 1 of 1)

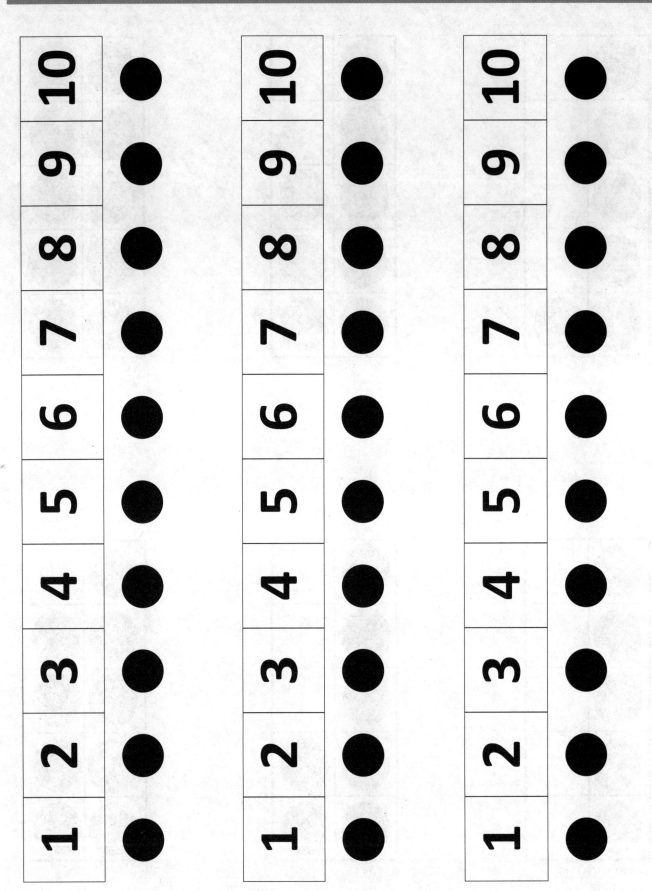

Activity 18 Hundreds Chart

1		3	4	5	6	7	8	9	10
11	12	13	14		16	17	18	19	20
21	22		24	25	26	27	28	29	30
31	32	33	34	35	36	37		39	40
41	42	43	44	45		47	48	49	50
51	52	53		55	56	57	58	59	60
61	62	63	64	65	66		68	69	70
71	72	73	74	75	76	77		79	80
81	82	83	84	85	86	87	88		90
91	92	93	94	95	96	97	98	99	

Activity 19 Hundreds Chart

1	2	3	4	5	6	7	8	9	
11	12	13	14	15	16		18	19	20
21	22	23	24	25		27	28	29	30
31	32		34	35	36	37	38	39	40
41		43	44	45	46	47	48	49	50
51	52	53	54	55	56	57		59	60
61	62	63	64	65	66	67	68		70
	72	73	74	75	76	77	78	79	80
81	82	83	84	85	86	87		89	90
91	92	93		95	96	97	98	99	100

Activity 20 Hundreds Chart

1	2	3	4	5	6	7	8	9	10
11	12	13		15	16	17	18	19	20
21	22	23	24	25		27	28	29	30
31	32	33	34	35	36	37	38		40

Activity 21 Hundreds Chart

1	2	3	4	5	6	7	8	9	10
11		13	14	15	16	17	18	19	20
21	22	23	24	25	26	27		29	30
31	32	33	34		36	37	38	39	40

Activity 22 Hundreds Chart (page 1 of 1)

1	2	3	4	5	6	7	8	9	10
11	12	13	14	15	16	17	18		20
21	22		24	25	26	27	28	29	30
31	32	33	34	35		37	38	39	40

Activity 23 Hundreds Chart

	2	3	4	5	6	7	8	9	10
11	12	13	14		16	17	18	19	20
21	22		24	25	26	27	28	29	30
31	32	33	34	35	36	37	38		40
41		43	44	45	46	47	48	49	50
51	52	53	54	55	56	57		59	60
61	62	63		65	66	67	68	69	70
71	72	73	74	75		77	78	79	80
81	82	83	84	85	86		88	89	90
91	92	93	94	95	96	97	98	99	

Activity 24 Hundreds Chart

1	2	3	4	5	6		8	9	10
11	12		14	15	16	17	18	19	20
21	22	23	24		26	27	28	29	30
31	32	33	34	35	36		38	39	40
41	42	43		45	46	47	48	49	50
51		53	54	55	56	57	58	59	60
61	62	63	64	65		67	68	69	70
71	72	73	74	75	76	77	78		80
81	82	83	84	85	86	87		89	90
91	92	93		95	96	97	98	99	**100**

Index

Page numbers followed by *c* and *f* indicate chart and figure, respectively.

Achievement outcomes, 1
Addition, 2, 8
   see also specific activities
Assessment of number sense, 2–3, 5
Automaticity, *see* Finger automaticity activities

Base ten activities, 5–7, 6*f*
Before/after activities
   Lesson 4, 38
   Lesson 5, 43
   Lesson 6, 48–49
   Lesson 7, 54
   Lesson 8, 59
   Lesson 9, 64
   Lesson 10, 69
   Lesson 12, 80
   Lesson 14, 90
   Lesson 16, 98
   Lesson 18, 106
   Lesson 20, 116
   Lesson 22, 126
   Lesson 24, 135
   lesson guidelines, 4, 5
   materials, 14, 182
Behavioral boundaries, 17
Bigger/smaller activities
   Lesson 11, 75
   Lesson 13, 85–86
   Lesson 15, 94
   Lesson 17, 102
   Lesson 19, 111
   Lesson 21, 121
   Lesson 23, 130
   lesson guidelines, 4, 5
   materials, 13, 182, 223
   see also Comparison of quantities activities

Cardinality activities
   Lesson 1, 18
   Lesson 2, 24
   Lesson 3, 31
   Lesson 4, 37
   Lesson 6, 48
   Lesson 7, 53
   Lesson 8, 58
   Lesson 9, 64
   Lesson 10, 69
   lesson guidelines, 4, 5
   materials, 182
   overview of, 6

Common Core State Standards (CCSS), 1, 5–8, 9–12*c*
Commutative property and partners, 7
Comparison of quantities activities
   Lesson 5, 44
   Lesson 6, 50
   Lesson 7, 55
   Lesson 8, 60
   Lesson 9, 65
   Lesson 10, 71
   materials, 183–189, 225
   see also Bigger/smaller activities
Counting, 2, 8, 112–113
   see also Counting warm-ups; Finger counting activities
Counting warm-ups
   Lesson 1, 17–18
   Lesson 2, 23
   Lesson 3, 30
   Lesson 4, 37
   Lesson 5, 42
   Lesson 6, 47
   Lesson 7, 53
   Lesson 8, 58
   Lesson 9, 63
   Lesson 10, 68
   Lesson 11, 73
   Lesson 12, 79
   Lesson 13, 84
   Lesson 14, 88
   Lesson 15, 92
   Lesson 16, 96
   Lesson 17, 101
   Lesson 18, 104
   Lesson 19, 109
   Lesson 20, 115
   Lesson 21, 119
   Lesson 22, 124
   Lesson 23, 129
   Lesson 24, 133
   materials, 211
   overview of, 5
Counting-all strategy, 8
Counting-on strategy, 8, 112–113
   see also Number operation activities

Dime activities, 14, 135, 213
Drawing story problems, 99–100, 103–104, 107–108, 176–178

Equals sign/quantities, 20, 44
Equations, *see* Number sentence activities
Error correction, 4, 18, 19

Error-free practice, 4
Extension partner activities, 137–144

Finger automaticity activities
    Lesson 1, 19–20
    Lesson 2, 25
    Lesson 3, 32
    Lesson 4, 39
    Lesson 5, 43–44
    Lesson 6, 49
    Lesson 7, 54
    Lesson 8, 59
    Lesson 9, 65
    Lesson 10, 70
    Lesson 11, 76
    Lesson 12, 81
    Lesson 13, 87
    Lesson 14, 91
    Lesson 15, 95
    Lesson 16, 98
    Lesson 17, 102
    Lesson 18, 107
    Lesson 19, 111
    Lesson 20, 117
    Lesson 21, 122
    Lesson 22, 126–127
    Lesson 23, 131
    Lesson 24, 135
    materials, 14
Finger calculating activities, 8
Finger counting activities
    Lesson 1, 19
    Lesson 2, 25
    Lesson 3, 32
    Lesson 4, 38–39
    Lesson 6, 49
    Lesson 7, 54
    Lesson 8, 59
    Lesson 9, 65
    Lesson 10, 70
    styles of, 7
Fingers and number sentences, *see* Number sentences on fingers
Five Frame activities, 14, 45, 193–209, 212

Gestures guidelines, 4
Goals for lessons, 9–12*c*

Hundreds chart activities
    Lesson 18, 105
    Lesson 19, 110
    Lesson 20, 115
    Lesson 21, 120
    Lesson 22, 125
    Lesson 23, 129
    Lesson 24, 134
    materials, 227–233

Learning goals, 9–12*c*
Lesson(s)
    activities by goals and standards, 5–8, 9–12*c*
    in daily classroom life, 8, 13

    guidelines for, 3–5
    research support for, 1, 3
Linear representations, 4

Magic Number activities
    Lesson 1, 18–19
    Lesson 2, 24–25
    Lesson 3, 31–32
    Lesson 4, 37–39
    Lesson 5, 42–43
    Lesson 6, 48–49
    Lesson 7, 53–54
    Lesson 8, 58–59
    Lesson 9, 64–65
    Lesson 10, 69–70
    Lesson 11, 74–75
    Lesson 12, 79–80
    Lesson 13, 84–85
    Lesson 14, 89–90
    Lesson 15, 93–94
    Lesson 16, 97–98
    Lesson 17, 101–102
    Lesson 18, 105–106
    Lesson 19, 110–111
    Lesson 20, 115–116
    Lesson 21, 120
    Lesson 22, 125–126
    Lesson 23, 129
    Lesson 24, 134–135
    lesson guidelines, 4, 5
    materials, 13–14, 182, 212–222
    overview of, 5–7
Making the number activities
    Lesson 8, 60
    Lesson 10, 70
    Lesson 11, 74–75
    Lesson 12, 79–80
    Lesson 13, 84, 87
    Lesson 14, 89
    Lesson 15, 93
    Lesson 16, 97
    Lesson 17, 101
    Lesson 18, 105
    Lesson 19, 110
    Lesson 20, 115–116
    Lesson 21, 120
    Lesson 22, 125
    Lesson 23, 129–130
    Lesson 24, 134
    materials, 13–14, 213, 214–222
Materials for lessons
    gathering/making, 13–15
    guidelines for, 4, 5
    see also specific activities
-1 activities
    Lesson 13, 86
    Lesson 14, 90
    Lesson 15, 94–95
    Lesson 16, 98
    Lesson 18, 106
    Lesson 20, 116
    Lesson 21, 121
    Lesson 22, 126
    Lesson 23, 131
    Lesson 24, 135

materials, 13–14, 211, 220–222, 224, 226
Minus sign, 21
Money activities, *see* Nickels/pennies activities

N – 1, *see* –1 activities
N + 1, *see* +1 activities
Nickels/pennies activities
    Lesson 19, 112
    Lesson 20, 117
    Lesson 21, 122
    Lesson 22, 127
    Lesson 23, 131–132
    Lesson 24, 135
    materials, 14, 213
NSS, see Number Sense Screener
Number knowledge activities, 1, 2
    see also specific activities
Number list activities
    Lesson 11, 76
    Lesson 12, 81
    Lesson 13, 86
    Lesson 14, 90
    Lesson 15, 94–95
    Lesson 16, 98
    Lesson 18, 106
    Lesson 20, 116
    Lesson 21, 121
    Lesson 22, 126
    Lesson 23, 131
    Lesson 24, 135
    lesson guidelines, 4, 5
    materials, 13–14, 211, 220–222, 224, 226
Number operation activities
    Lesson 19, 112–113
    Lesson 20, 117
    Lesson 21, 122
    Lesson 22, 127
    Lesson 23, 132
    Lesson 24, 136
    materials, 14
    overview of, 1, 2
Number recognition activities, 5–7, 6*f*
    see also Magic Number activities
Number relations activities, 1, 2
    see also specific activities
Number sense, 1–3, 5
Number sense interventions
    in daily classroom life, 8, 13
    by goals and standards, 5–8
    guidelines for, 3–5
    research support for, 1, 3
Number Sense Interventions Activities Organized by Learning Goals with Common Core Framing, 9–12*c*
Number Sense Screener (NSS), 2–3
Number sentence activities
    Lesson 19, 112–113
    Lesson 20, 117
    Lesson 21, 122
    Lesson 22, 127
    Lesson 23, 132
    Lesson 24, 136
    materials, 14
    story problems and, 8
    see also Number sentences on fingers; Partners and number sentences

Number sentences on fingers
    Lesson 1, 22
    Lesson 2, 28–29
    Lesson 3, 35
    Lesson 4, 41
    Lesson 5, 45–46
    Lesson 6, 51
    Lesson 7, 56
    Lesson 8, 62
    Lesson 9, 67
    Lesson 10, 72
    materials, 13, 193–209, 210
    see also Number sentence activities

Operations, *see* Number operation activities
Oral counting, *see* Counting warm-ups

Pacing and progress, 5
Partner set activities
    extension activities, 137–144
    Lesson 1, 20–21
    Lesson 2, 26–28
    Lesson 3, 33–34
    Lesson 4, 40
    Lesson 5, 44–45
    lesson guidelines, 4
    materials, 13–15, 190–210 , 212–213
    overview of, 2, 7
Partners and number sentences
    Lesson 1, 20
    Lesson 2, 26–27
    Lesson 3, 33–34
    Lesson 4, 40
    materials, 13, 190, 193–210
    overview of, 7
Pennies activities, *see* Nickels/pennies activities
+1 activities
    counting on and, 8
    Lesson 11, 76
    Lesson 12, 81
    Lesson 15, 94–95
    Lesson 16, 98
    Lesson 18, 106
    Lesson 20, 116
    Lesson 21, 121
    Lesson 22, 126
    Lesson 23, 131
    Lesson 24, 135
    lesson guidelines, 4, 5
    materials, 13–14, 211, 220–222, 224, 226
Plus sign, 20
Progress and pacing, 5

Recognizing numbers, 5–7, 6*f*
    see also Magic Number activities
Recognizing sets activities
    Lesson 1, 19–20
    Lesson 2, 25–26
    Lesson 3, 32
    Lesson 4, 39
    Lesson 5, 44
    Lesson 6, 50
    Lesson 7, 54

Recognizing sets activities—*continued*
- Lesson 8, 59
- Lesson 9, 65
- Lesson 10, 71
- Lesson 12, 82
- Lesson 13, 87
- Lesson 14, 91
- Lesson 15, 95
- Lesson 16, 98
- Lesson 17, 103
- Lesson 18, 107
- materials, 7, 13, 183–189, 225

Seating arrangements, 3
Sequencing and number recognition activities
- Lesson 1, 18–19
- Lesson 2, 24–25
- Lesson 3, 31
- Lesson 4, 37–38
- Lesson 5, 42–43
- Lesson 6, 48
- Lesson 7, 53–54
- Lesson 8, 58
- Lesson 9, 64
- Lesson 10, 69
- Lesson 11, 75
- Lesson 12, 80
- Lesson 13, 85
- Lesson 14, 90
- Lesson 15, 94
- Lesson 16, 97
- Lesson 17, 102
- Lesson 18, 106
- Lesson 19, 111
- Lesson 20, 116
- Lesson 21, 120
- Lesson 22, 125–126
- Lesson 23, 130
- Lesson 24, 134
- materials, 14
- overview of, 6

Sets, *see* Partner set activities; Recognizing sets activities
Smaller/bigger activities, *see* Bigger/smaller activities
Standards, 1, 5–8, 9–12*c*
Story problems
- Lesson 1, 21–22
- Lesson 2, 28
- Lesson 3, 34–35
- Lesson 4, 41
- Lesson 6, 50–51
- Lesson 7, 55–56
- Lesson 8, 61
- Lesson 9, 66–67
- Lesson 10, 71–72
- Lesson 11, 77
- Lesson 12, 82
- Lesson 16, 99–100
- Lesson 17, 103–104
- Lesson 18, 107–108
- Lesson 19, 113–114
- Lesson 20, 118
- Lesson 21, 123
- Lesson 22, 127–128
- Lesson 23, 132

Lesson 24, 136–137
- materials, 13–15, 176–179, 193–209, 212–213
- number sentences and, 8
- overview of, 7, 8

Subitizing activities
- Lesson 1, 19–20
- Lesson 2, 25–26
- Lesson 3, 32
- Lesson 4, 39
- Lesson 5, 43–44
- Lesson 6, 49–50
- Lesson 7, 54–55
- Lesson 8, 59
- Lesson 9, 65
- Lesson 10, 70
- Lesson 11, 76
- Lesson 12, 81–82
- Lesson 13, 87
- Lesson 14, 91
- Lesson 15, 95
- Lesson 16, 98–99
- Lesson 17, 102–103
- Lesson 18, 107
- Lesson 19, 111–112
- Lesson 20, 117
- Lesson 21, 122
- Lesson 22, 126–127
- Lesson 23, 131–132
- Lesson 24, 135–136
- materials, 7, 13–14, 183–189, 213, 225
- overview of, 7, 8

Subtraction, 2, 118
see also specific activities

Ten Frame activities
- Lesson 8, 60–61
- Lesson 9, 66
- Lesson 10, 70–71
- Lesson 11, 76–77
- materials, 7, 14–15, 213

Vocabulary guidelines, 5

Warm-ups, *see* Counting warm-ups
Written numbers activities
- importance of, 4
- Lesson 1, 22
- Lesson 2, 29
- Lesson 3, 36
- Lesson 4, 41
- Lesson 5, 46
- Lesson 6, 52
- Lesson 7, 57
- Lesson 8, 62
- Lesson 9, 67
- Lesson 10, 72
- Lesson 11, 78
- Lesson 12, 83
- Lesson 13, 87
- Lesson 14, 91
- Lesson 15, 95
- materials, 146–175
- overview of, 7